AF646390

LEARN TO GROW MUSHROOM

The Author

Dr. Prabhat Kennedy Soreng S.J. comes from Simdega district of the State of Jharkhand. He did his high schooling from St. Mary's High School Samtoli, Intermediate Science from St. Xavier's College Ranchi, BSc from Loyola College Chennai, and MSc Botany with Microbiology as a special paper from St. Joseph College Bangalore. He has qualified UGC-NTA and UGC-CSIR NET. Philosophical studies from Chennai (Madras) University. He did Religious studies from Pune and PhD from Ranchi University Ranchi. His PhD thesis was- Cultivation of Wild Edible Mushrooms of Jharkhand. At present he is an assistant professor in the Department of Botany, St. Xavier's College Ranchi.

LEARN TO GROW MUSHROOM

Author

Dr. Prabhat Kennedy Soreng S.J.

2025

Daya Publishing House®

A Division of

Astral International Pvt. Ltd.

New Delhi - 110002

© 2025 AUTHOR

ISBN: 978-93-5919-768-5

Publisher's Note:

Every possible effort has been made to ensure that the information contained in this book is accurate at the time of going to press, and the publisher and author cannot accept responsibility for any errors or omissions, however caused. No responsibility for loss or damage occasioned to any person acting, or refraining from action, as a result of the material in this publication can be accepted by the editor, the publisher or the author. The Publisher is not associated with any product or vendor mentioned in the book. The contents of this work are intended to further general scientific research, understanding and discussion only. Readers should consult with a specialist where appropriate.

Every effort has been made to trace the owners of copyright material used in this book, if any. The author and the publisher will be grateful for any omission brought to their notice for acknowledgement in the future editions of the book.

All Rights reserved under International Copyright Conventions. No part of this publication may be reproduced, stored in a retrieval system, or transmitted in any form or by any means, electronic, mechanical, photocopying, recording or otherwise without the prior written consent of the publisher and the copyright owner.

Published by : **Daya Publishing House®**
A Division of
Astral International Pvt. Ltd.
– ISO 9001:2015 Certified Company –
4736/23, Ansari Road, Darya Ganj
New Delhi-110 002
Ph. 011-43549197, 23278134
e-mail: info@astralint.com
Website: www.astralint.com

Printed at : **Replika Press Pvt Ltd**

Message from Fr. Principal

Dr. Fr. Nabor Lakra s.j.

Principal

St. Xavier's College Ranchi

Books are the sources of knowledge. They are written by the scholars to share their knowledge. I am pleased that Dr. Prabhat Kennedy Soreng, assistant professor in the Department of Botany and Registrar of the College publishing the book, "Learn to Grow Mushroom".

In the CBCS undergraduate syllabus of Botany, mushroom culture technology is one of the Skill Enhancement Course and in the NEP syllabus mushroom cultivation is one of the vocational papers. This book is written to help those students who would be studying these papers and anyone who is interested to learn to grow mushroom in his/her own house. This book also focuses on the mushrooms of Jharkhand.

I hope that this book will be of great help to the students.

Message from HOD

Dr. Ajay Kumar Srivastava

HOD, Botany
St. Xavier's College Ranchi

With rainfall coming in one season only and little assured irrigation, Jharkhand farmers are thrown unemployed for many months in the year. More so with the tradition of monoculture (rice) the spectre of malnutrition in very apparent. But in the wilderness Jharkhand boasts of not only tree diversity but also edible mushroom diversity. The people around the forest make some food and money from these but the basic problem remains nearly the4 same.

The effort of the author Fr. Prabhat K Soreng s.j is to acquaint the common man with mushroom cultivation so that they could get protein rich food source while being employed round the year apart from maintaining the traditional knowledge.

This book intends to educate the students from various disciplines to get engaged part time and augment the family income.

Preface

My parents had a very good knowledge of the wild edible mushrooms though they were illiterate. They used to collect a lot of edible mushrooms from the forest in the beginning of the rainy season for consumption. I got some knowledge about mushrooms from them and am able to identify wild edible mushrooms especially of Jharkhand.

Now, as an Assistant Professor in the Department of Botany, St. Xavier's College, Ranchi I have a special interest in mushrooms because they have beautiful and attractive looks of their fruiting bodies and mushrooms are one of the key forest products gathered by the tribal people of Chhotanagpur especially in the rainy season. Jharkhand is gifted with rich diversity of mushrooms, both edible and poisonous.

Now and then news appears in the news papers about people dying or falling sick due to consumption of poisonous mushrooms. This also inspired me to study about wild mushrooms, after studying about mushrooms I came to know that they are rich sources of proteins and other nutrients yet I did not find any short cut method to distinguish edible and poisonous mushrooms. Only after experience we can single out the edible mushrooms from others. Now with the advancement of technology it may be possible to differentiate edible from poisonous mushrooms. Since the tribal people are very close to the nature in Jharkhand and depend on forest products, to some extent they have good knowledge of wild edible mushrooms. With the help of them we can identify edible mushrooms. In this book, I have tried to illustrate the mushrooms with pictures and have named and classified those using different criteria.

In this book, I have also written the steps, procedure and requirements for growing mushrooms in the culture lab or at home. This book will be very helpful for those students who want to learn to grow mushrooms and want to identify local mushrooms especially mushrooms from Jharkhand.

I am hopeful that this book will be of great help to students and those who would like to know about mushrooms and want to learn to grow them in the culture lab or at home.

Dr. Prabhat Kennedy Soreng

Assistant Professor,

Department of Botany

St. Xavier's College, Ranchi

Acknowledgment

Attitude is the gratitude of heart. There are people who have been source of inspiration and encouragement to me to publish this book; I am indebted and grateful to them.

I thank the Almighty for giving me strength to do research work and gather some knowledge about mushrooms.

I will be ever grateful to my loving parents (John Soreng and Anna Soreng) who gave me basic knowledge about wild edible mushrooms.

Sincere thanks to Dr. Ajay Kumar Srivastava, Head of the Department of Botany, St. Xavier's College, Ranchi for fruitful suggestions and guidance.

I am thankful to all faculty members of the Department of Botany, St. Xavier's College, Ranchi for their encouragement and support.

I am also grateful to the Principal of St. Xavier's College, Ranchi and to my Authorities for giving me space and time to publish this book.

I am extremely grateful to the scholars whose resource materials; I have referred and used especially Google images and Wikimedia creative commons images.

Finally thanks to all my well wishers especially to my dear students for their encouragement.

Dr. Prabhat Kennedy Soreng S.J.

Acknowledgement

Contents

Chapter 1

Introduction

What are Mushrooms?

Mushrooms are fleshy and macroscopic *Basidiocarpous* or *Ascocarpous* spore bearing fungi. They are attractive because of their beauty and nutritive value though there are many mushrooms which are deadly poisonous. There are thousands of mushrooms species on earth. Some of them have been identified as edible and have been domesticated and they are grown in a large scale for the commercial purpose.

Mushrooms are unique; they are different from animals and plants. They lack the green chlorophyll that plants use to synthesise their own food and energy. For this reason, they are placed in the Kingdom of Fungi. Mushrooms are also unique within the Fungal Kingdom itself, because they produce the complex fruiting body which we call as, 'The Mushroom', all the mushrooms are placed in a division called 'Eumycota' meaning 'The True Fungi'.

Ainsworth *et al.* grouped mushrooms into two sub divisions. The two sub divisions are *Basidiomycotina* and *Ascomycotina*. The latter, produces ascospores in ascus during sexual reproduction while the members of the *Basidiomycota* produce basidiospores in the basidium.

Mushrooms are conspicuous umbrella-shaped fruiting bodies, typically of order *Agaricales* in the class *Basidiomyctes*. A mushroom is the fruit of the fungal organism that produces them, just like mango tree produces mango to bear seed to ensure the continuation of its species, so the fungal organism produces mushrooms that carry spores to ensure the continuation of its own species. Therefore, the mushroom is the reproductive organ for the fungus. This means that by picking a mushroom we do not harm the fungus itself, because the main body of the organism lies underground in the form of a network of minute threads called 'Hyphae'. When two compatible hyphae meet they join together to form another network called the 'Mycelium' which grows quietly and unseen underground for most of the year until the conditions are right for fruiting and that's when we get to see mushrooms.

Mushrooms are very delicate things, they have a life span of less than a day others may survive one week, and a group of tougher mushrooms may last months but they have a tough woody texture. Most fleshy mushrooms do not last, and this makes researching them very difficult. Since we do not know where they are till they fruit, we only get a few days to study them each time, and they are seasonal.

Each Mushroom carries within it millions or even billions of spores, to the extent, that in the case of some kinds of mushrooms grown commercially, workers have to wear dust masks to protect themselves from the spore dust and breath easily. Only a few of these spores manage to survive and grow into a mycelia network producing new mushrooms. To make life even more difficult, two compatible spores have to meet to be able to produce mushrooms.

Mushroom classification and identification is a difficult task because of lack of easier keys. Mushrooms have been classified using different parameters like morphology (shape, size, colour), structure and texture of fruiting bodies, natural habitat, occurrence, spore colour, edibility etc

Chapter

2

Morphology of A Mushroom

A mushroom has mainly two portions, one is fruiting body (sporocarps) which is usually umbrella shape and second is mycelium which is thread like structure grows on the substrate. Mycelium can be compared to the roots of higher plants. Mycelium helps in absorbing the nutrients from the substrate. Most of the mushrooms have caps and stalks but some are devoid of stalks. In addition, not all members of the *Basidiomycota* are mushrooms with stalk, caps and gills. There are coral fungi, polypores, puffballs and boletes, just to name a few.

Different Parts of a Typical Mushroom

The stalk of the mushroom bears a ring; it is also known as the annulus. Most mushrooms don't have rings but some do. The presence or absence of a ring is one of the first things you need to notice when identifying edible or poisonous mushrooms. The same is true about the cuplike affairs known as cups or volvas, which is present at the mushroom stalk's base. We need to brush away leaves in order to be sure whether mushroom does or does not arise from a cup especially when we study them in their natural habitats. Sometimes the mushroom's stalk, instead of arising at the center of the cap, attaches to the side of the cap. Many mushrooms, such as the *Boletus edulis* Bull., instead of having *gills* radiating from the central stalk, have roundish holes, or pores.

While identifying mushrooms there are many unique and interesting features to use as field marks. Any good mushroom field guide will present a brief survey of mushroom parts, with drawings showing how the parts can vary, and what the various configurations and states are called.

A mushroom may have following parts but some mushrooms may not contain all these structures.

- **The caps or pileus:** This is the most attractive part of a mushroom. This could be thick, fleshy, membranous or corky depending on the variety of mushrooms. Pileus surface may be smooth, rough, scaly or hairy. The apex of the cap could be convex, conical or campanulate.

- **The gills or lamellae:** they are situated below the pileus and extend from the tip of the stalk to the margin of the pileus. Basidia develop in the gills which produce spores.
- **The veil:** It covers the gills and extents from the margin of the cap to the stalk. When fruiting body matures it expands along the cap and breaks away.
- **The volva:** Some varieties of mushrooms remains covered by a veil, when fruiting body develops this veil breaks open and some parts of it remain attached to the stalk forming cup like structure called as volva.
- **The annulus or ring:** It is formed around the stipe. It may be present or absent.
- **The stipe:** It is centrally attached to the cap. In some mushrooms it may be attached laterally to the cap. It may be solid or hollow. It may be present or absent in many species. Its attachment with pileus could be central, eccentric or lateral. These parts are helpful for mushroom identification and their classification.

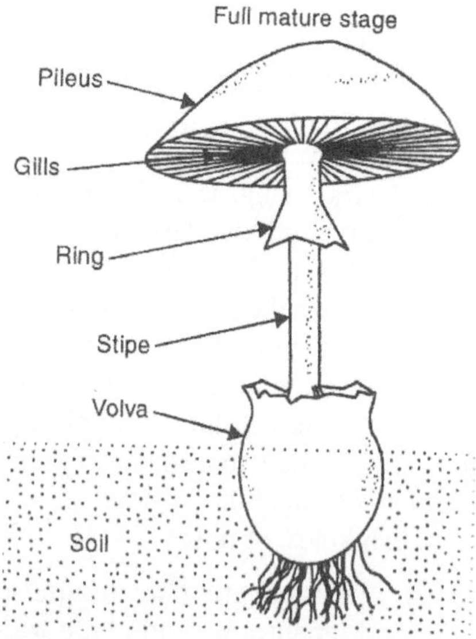

Fig. 1. Parts of a typical mushroom

Life-Cycle of a Mushroom

Fruiting bodies of mushrooms are parts of the sexual reproductive cycle. Deacon states that the sexual reproduction must be regarded as a whole sequence of events each with its own developmental features: (I) production of sex organs and gametes; (II) fusion of gametes (plasmogamy) followed by nuclear fusion (karyogamy); (III) meiosis and subsequent nuclear division; (IV) Development of fruiting bodies to disperse the sexual spores; and (V) development of spores.

The life cycle of a mushroom starts when the spores are released from the gills, (whichever surface the mushroom happens to carry spores on). Millions of spores are released into the elements, (air, water, animals) these spores are dispersed by various methods, (depending on the kinds of mushrooms).

When the conditions are right, the spores germinate sending out tiny threads called hyphae (single: hypha, monokaryotic). In order to develop and eventually produce a mushroom it has to find other hyphae that are compatible. When two compatible hyphae meet, they fuse together to form a network or threads called a mycelium (dikaryotic). This mycelium eventually forms what is known as a hyphal knot which grows and develops into a pinhead which in turn grows and develops into a fruiting body what we call mushroom and then it all starts again. Variation in the timing, sequence, mechanism and regulation of the basic steps yields the incredibly varied life cycles encountered within the fungi. The knowledge of the life cycle of a mushroom is useful for the mushrooms cultivators.

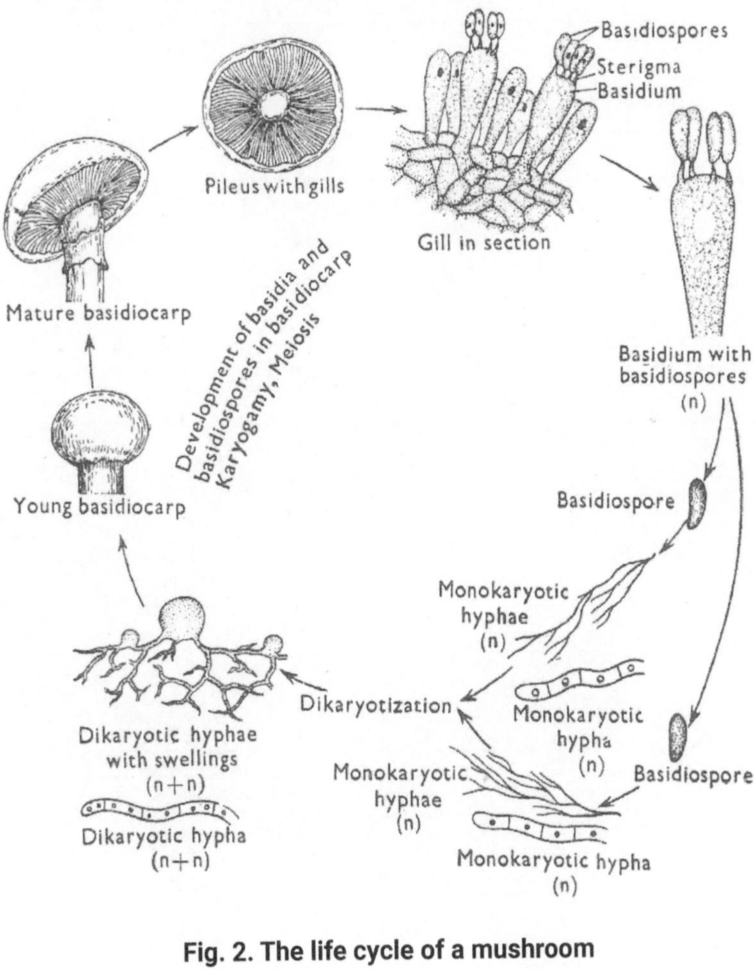

Fig. 2. The life cycle of a mushroom

Substrates and Habitats of Mushrooms

Mushrooms are very specific to their nutrient source thus they grow on a particular habitat for their nutrient requirement. They grow saprophytically on various dead substrates and parasitically or symbiotically on or with other organisms. Their fruiting bodies occur either above the substratum or hidden under the substratum. Mushrooms can be classified into three main groups based on the natural habitat. Humicolous (Humus inhabiting), Lignicolous (Wood inhabiting) and Coprophilous (Dung inhabiting). The natural habitat of mushrooms is helpful to know the nutrient requirement of a particular group of mushrooms and their chemical composition.

Chapter 3

Classification of Mushrooms

Mushrooms can be classified into different groups using different criteria, like, morphology, occurrence, natural habitat, Structure and Texture of Fruit Bodies, colour of spores, edibility. All these criteria are used to classify them and they are useful for mushroom identification.

Based on Morphology

Table 1. Classification based on morphology

Basidiomycetous (spores formed in the basidium)	Fruiting layers exposed to the air e.g. gill fungi, pore fungi, teeth fungi, coral fungi or jelly fungi.
	Fruiting layers not exposed to the air e.g. *Dictyophora*, *Ithyphallus*, *Geaster*, *Lycoperdon*, *Bovista*, *Calvatia*, *Cyathus*
Ascomycetous (spores formed in the ascus)	Mushrooms with pitted or gnarled cap e.g. *Morchella*, *Gyromitra*,
	Mushrooms with saddled-shaped cap e.g. *Helvella*
	Mushrooms saucer, cup or globlet-shaped e.g. *Plicaria*, *Pustularia*, *Sarcoscypha*, *Geopyxis*

Gill fungi
rawpixel.com

Pore fungi
commons.wikimedia.org

Teeth fungi
commons.wikimedia.org

Club or Coral
picryl.com

Jelly fungi
commons.wikimedia.org

Dictyophora
pixabay.com

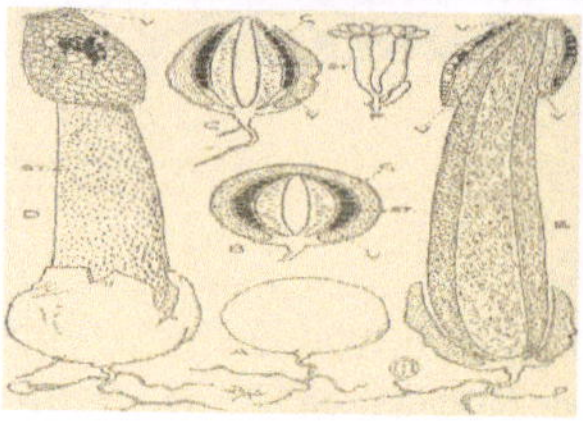

Ithyphallus garystockbridge 617.
getarchive.net

Geaster
www.flickr.com

Lycoperdon
commons.wikimedia.org

Bovista
commons.wikimedia.org

Calvatia
commons.wikimedia.org

Cyathus
commons.wikimedia.org

Morchella
commons.wikimedia.org

Gyromitra
commons.wikimedia.org

Helvella
commons.wikimedia.org

Plicaria
commons.wikimedia.org

Sarcoscypha
commons.wikimedia.org

Geopyxis
commons.wikimedia.org

Fig.3. Examples

Based on Occurrence

Mushrooms, the fruiting bodies occur either above the substratum or hidden under the substratum.

Epigenous	Fruiting bodies are formed above the surface of substratum e.g. *Pizza* sp. *Morchella* sp.
Hypogenous	Fruiting bodies are formed underground e.g. *truffles*

Cup Fungi, *Pizza* sp.
pxhere.com

Fig. 4. Cup Fungi, Pizza sp.

Morel, *Morchella* sp.
commons.wikimedia.org

Fig. 5. Morels, *Morchella* sp.

Based on Natural Habitat

Mushrooms are very specific to their nutrient source thus they grow on particular habitat for their nutrient requirement.

Table 2. Classification based on Natural habitat

Humicolous	Humus inhabiting
	Saprophytic e.g. *Lepista nuda*, *Volvariella*, *Marasmium*, *Polyporus tuberaster*
	Symbiotic e.g. *Boletus*, *Lacterius*, *Tricholoma*, *Termitomyces*
Lignicolous	Wood inhabiting
	Saprophytic e.g. *Agrocybe*, *Pleurotus*, *Auricularia*, *Lentinus*
	Parasitic, e.g. *Armellariella mellea*, *Cyttaria*
Coprophilous	Dung inhabiting, e.g. *Agaricus*, *Caprinus*

Lepista nuda
commons.wikimedia.org

Volvariella
commons.wikimedia.org

Marasmium

commons.wikimedia.org

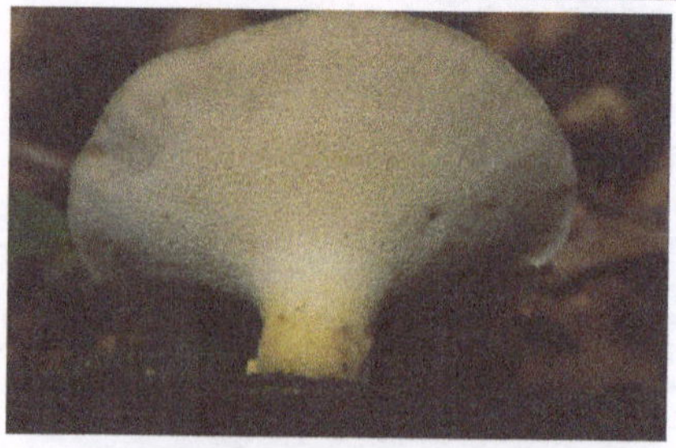

Polyporus tuberaster

commons.wikimedia.org

Boletus

www.rawpixel.com

Lacterius

commons.wikimedia.org

Tricholoma

commons.wikimedia.org

Termitomycetes

commons.wikimedia.org

Agrocybe

commons.wikimedia.org

Pleurotus

commons.wikimedia.org

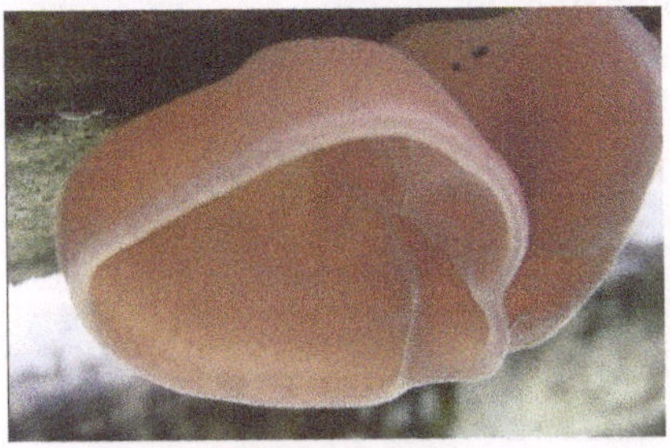

Auricularia

commons.wikimedia.org

Lentinus

commons.wikimedia.org

Armellariella mellea

commons.wikimedia.org

Cyttaria

commons.wikimedia.org

Agaricus

www.publicdomainpictures.net

Caprinus

commons.wikimedia.org

Fig.6. Examples

Based on Colour of Spores

Out of several criteria used for mushroom identification spore print is one of them. Mushrooms produce spores of different colours like white, yellow brown, pink, purple brown, black which are the genetic features.

The spore print is a simple method used for identification of gilled fungi. When the cap mushroom is held stationary on a paper in a closed chamber for a long period, the impression of pattern of gills is printed, and the spores get accumulated onto the paper and invisible spores appear coloured.

Table 3. Colour of spores of mushrooms

Mushroom types	***Genus***	***Spore colour***
Gilled mushrooms:	*Amanita*	White to pale
	Chlorophylum	Green
	Clitocybe	Fleshy colour
	Hebeloma	Rusty to yellow brown
	Hygropharus	Cream
	Inocybe	Brown or cinnamon
	Lepiota	Green
	Naematoloma	Purple brown
	Panaeolus	Black
	Pleurotus	Lavendor colour
	Russula	White, cream or yellow
	Schizophyllum	Pink, salmon or flesh
Tubed mushrooms:	*Boletus*	Yellow to olive brown
	Cyrodon	Yellow to yellow brown
	Suillus	Yellow brown, cinnamon
Puffball mushrooms:	*Scleroderma*	Purple black

Based on the colour of the spores, mushrooms can be classified into five groups.

Table 4. Classification based on spore colour

White Spored Mushrooms 	E.g. *Amanita,Armillaria, Cantharellus, Clitocybe, Lacterius, Lentinus, Lepiota, Pleurotus, Russula, Tricholoma,*
Yellow Brown Spored Mushrooms 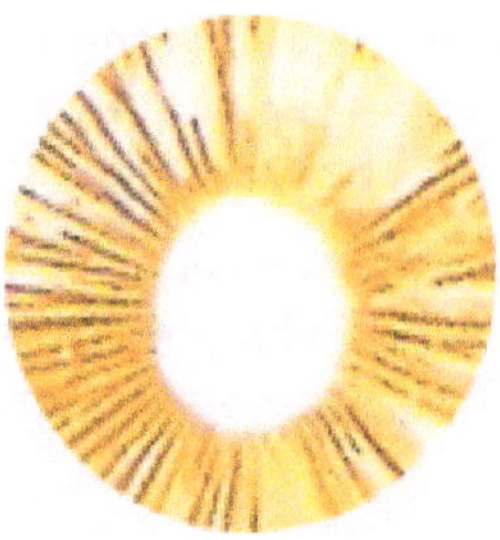	e.g. *Flammulina, Inocybe, Paxillus*

Pink Spored Mushrooms 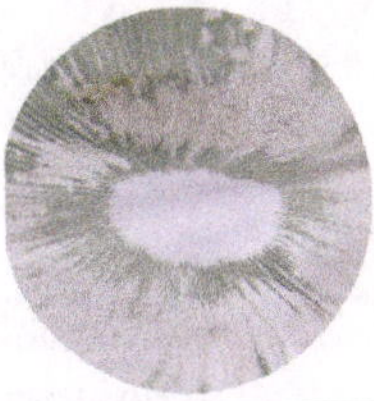	e.g. *Entoloma, Volvariella*
Purple Brown Spored Mushrooms 	e.g. *Agaricus, Hypholoma, Stropharia*
Black Spored Mushrooms 	e.g. *Panaeolus, Caprinus, Panaeolus*

 ***Amanita*** commons.wikimedia.org	 ***Cantharellus*** commons.wikimedia.org	 ***Clitocybe*** commons.wikimedia.org
 ***Lacterius*** commons.wikimedia.org	 ***Lentinus*** commons.wikimedia.org	 ***Lepiota*** commons.wikimedia.org

Pleurotus
commons.wikimedia.org

Russula
commons.wikimedia.org

Tricholoma
commons.wikimedia.org

Flammulina
commons.wikimedia.org

Inocybe
commons.wikimedia.org

Paxillus
commons.wikimedia.org

Entoloma
commons.wikimedia.org

Volvariella
commons.wikimedia.org

Agaricus
commons.wikimedia.org

Hypholoma
commons.wikimedia.org

Stropharia
commons.wikimedia.org

Caprinus
commons.wikimedia.org

Panaeolus
commons.wikimedia.org

Psathyrella
commons.wikimedia.org

Fig. 7. Examples

Based on Structure and Texture of Fruit Bodies

Based on the structure and texture of fruiting bodies, mushrooms can be classified into

Table 5. Classification based on structure and texture of fruiting bodies

Gilled Fungal Mushrooms	This group of mushrooms is characterized by the presence of gills or lamella which bears spores. Genus *Agaricus* and *Amanita* of family Agaricaceae are the examples.
Pore Fungal Mushrooms	The fruit bodies are tough, leathery, or woody and pore layer cannot be separated from the flesh. Members of family Polyporaceae like *Polyporus umbellatus, Fistulina, Trametes, Favolus tenuiculus, Polyporus parennis* belong to this group of mushrooms. Pore shape could be diamond, round, angular, irregular etc.
Tooth Fungal Mushrooms	Fruits bodies white, usually in a clump of several, one above the other, each shelf 10-30 cm wide on living hard maple. Genus *Hydnum* of family Hydnaceae is the example of Tooth fungal mushroom.
Boletaceous Fungal Mushrooms	Large fruiting bodies, they can be identified or distinguished among them only by microscopic characteristics. *Boletinus, Boletus* of Boletaceae family are the examples.
Club Shaped Mushrooms	Mushrooms repeated branched, usually growing on decaying wood. Genus *Clavaria* is the most common fleshy fungi of this type belonging to Clavariaceae family.
Jelly Fungal Mushrooms	They are jelly like especially when moist and most often found growing on wood. Fruit body is gelatinous. Examples, *Exidia grandulosa, Tremella lutescens.*
Cup Shaped Mushrooms	Morchella species of phylum Ascomycota are the common examples of cup shaped mushrooms. Other examples are *Verpa, Patella, Helvella* and *Xylaria.*
Puffballs and Its Relatives	Fruit bodies are without a stalk extending through the interior. Examples are *Lycoperdon, Calvatia, Geaster, Bovista*, these are called puffballs.

Amanita pantherina
commons.wikimedia.org

Favolus tenuiculus
commons.wikimedia.org

Hydnum umblicatum
commons.wikimedia.org

Boletinus cavipes

commons.wikimedia.org

Clavariadelphus pistillaris

commons.wikimedia.org

Exidia thuretiana

commons.wikimedia.org

Tremella

commons.wikimedia.org

Xylaria hypoxylon

commons.wikimedia.org

Calvatia

commons.wikimedia.org

Fig. 8. Examples

Chapter 4

Mushroom Culture Technology

References of use of mushrooms are very old but mushroom culture technology is very recent. Comprehensive method of mushroom cultivation was first recorded by Tournefort (1707). Chambray (1810) and Callow (1831) made mushroom cultivation possible throughout the year. Out of 2000 Known edible mushrooms only 20 species have been domesticated. Some of the domesticated and commercially grown mushrooms are Oyster mushroom (*Pleurotus species*), Button mushrooms (*Agaricus brunnescens or Agaricus bisporus*), paddy straw mushrooms (*Volvariella volvacea*), Shitake mushroom (*Lentinus edodes*), nameko (*Pholiata nameko*), enokitake (*Flammulina velutipes*), Jew's ear fungi (*Auricularia sp.*) etc.

Requirements for Mushroom Cultivation

If anyone wants to do business by growing and selling mushrooms following things are required.

For selecting a site for establishing a mushroom farm the existing environmental and agro-climatic condition of the area is important especially with relation to the types of mushrooms to be grown. Second important factor is availability of enough raw materials and cheap human power. Mushroom farm should have composting yard, pasteurization chamber, and rooms for the preparation of spawn, spawn running, cropping and post harvest handling unit. It should have clean water supply, 3-phase electricity connection and proper drainage system. The composting yard should be close to the road. Pasteurization chamber should be built in such a way that it should open nearer to the cropping room. Composting yard should have infrastructure for long method of composting.

Pasteurization chamber is equipped with single zone system (Peak Heating Chamber) or double zone system (Bulk pasteurization chamber/Tunnel). Spawn lab is provided with inoculation, incubation, boiling, autoclaving, washing and cold storage systems. Growing rooms are two types, natural environmental conditions growing room and environment controlled conditions growing rooms.

A farm infrastructure with composting platform, bulk chamber, growing room, water supply, substrates easily available,mushroom unit (thatched house),Laminar Air Flow chamber, Spirit lamp, 70% alcohol,Polythene bag, vessels, inoculation hook, autoclave, low cost stove, sieves, culture rack, water sprayer, tray,formalin, bouvistin, Calcium carbonate, Calcium sulphate

Steps in Mushroom Cultivation

Procedure for mushroom cultivation may differ based on types of mushrooms one wants to grow but in general following steps have to be followed.

1. Preparation of PDA medium and PDA slants

Potato tubers are taken; they are peeled off and weighed 200g. They are chopped into small pieces with the help of a knife. Chopped potato pieces are transferred into a beaker containing about 100 ml of distilled water. The content is boiled with the help of a heater for about 20 minutes. The supernatant is filtered with the help of the muslin cloth and the filtrate was collected into a beaker. This filtrate is potato extract. To this, extract Dextrose, Agar and Peptone are transferred and gently heated and stirred to dissolve the ingredients. This medium is transferred to a measuring cylinder of 1litre capacity and made the volume to 1litre by adding distilled water. The pH of this medium is adjusted to 5.6, 6.5 by using 1NHCl or 1NNaOH. Then the medium is poured into conical flasks and they are plugged with cotton and are autoclaved at 121°C for 20 minutes. When the temperature cools down, the flasks are taken out from the autoclave and the medium is used to prepare Agar slants for mycelia culture.

Table 6. Composition of PDA medium

Preparation culture of Media

Potato Dextrose Agar (PDA) medium

Composition	
Potato tubers	200g
Dextrose	20g
Agar	15g
Peptone	2g
Distilled water	1 llitre
HCl	1N
NaOH	1N
pH	5.6

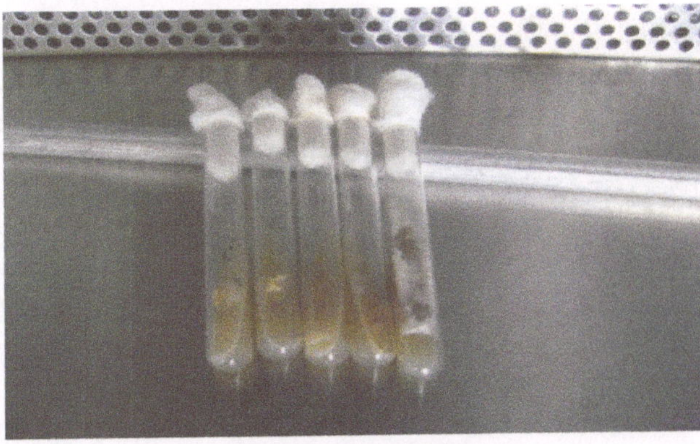

Fig. 9. Media slants

2. Collection of fresh mushrooms and Isolation of pure culture

For isolation and maintenance of pure cultures fresh mushrooms are brought to laboratory and used as starter material. Pure cultures are obtained from the pileus content by scooping a portion of the fleshy tissue and transplanting aseptically into PDA slants (Potato Dextrose Agar medium and incubated at room temperature under glass hoods for 7-15 days. These cultures are free from other contaminants. They are sub-cultured and stored in the refrigerator at 4°C and further used in the preparation of spawn for cultivation of mushrooms.

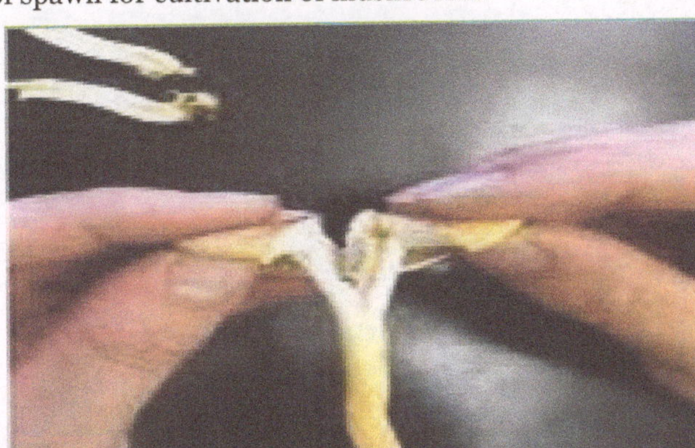

Fig. 10. Taking out the dikaryotic mycelia

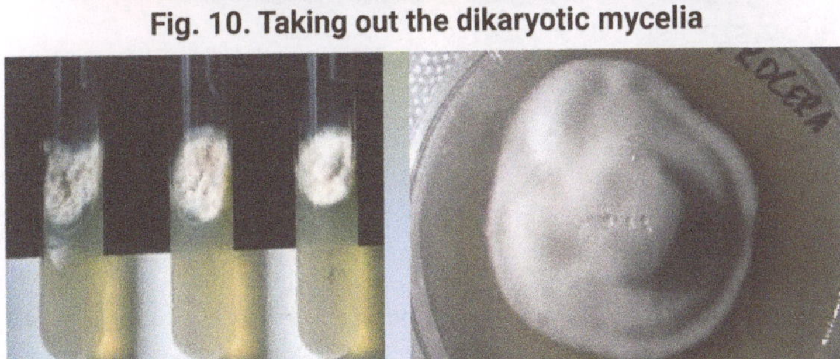

Fig. 11. Growth of mycelia on PDA slants

3. Spawn Preparation

Spawn is prepared on sorghum or wheat grains. Grains are weighed and boiled in 1litre of water in an open vessel on a hot plate till all the water evaporates. Occasionally, the contents are stirred using a wooden spatula and care is taken not to over boil or split the grains. Excess water in the grains is drained and spread on a tray. Gypsum added to remove the excess moisture and to prevent the grains from sticking to each other. Calcium Carbonate powder is added which is helpful to adjust a particular pH. These grains are then filled 2/3 into the 500ml. empty clean bottles or polythene bags and plugged with cotton. It is then autoclaved at 121° C for 30 minutes for three consecutive days. When, it is cooled, a small portion of the pure culture of the mycelia of the fungus is introduced into a few bottles and incubated at room temperature.

After two weeks of incubation, the grains are fully colonized by the fungal mycelia. Initially the bottle is often shaken to increase the incubation points. This spawn bottle is further used to inoculate other bottles (spawn multiplication). Contaminated bottles are removed and only pure spawn bottles are used for further cultivation work.

Table 7. Proportion of $CaCO_3$ and $CaSO_4$

SPWAN PREPARATION

WHEAT GRAIN	1 Kg
CaCO3	6gm
GYPSUM-Ca SO4	12gm
250-300 gm in the bottle or polythene bag	
10 kg Straw	
Water	100 litre
Bovistin	10gm
Formaline	100 ml
Humidity content of the straw	68-70%

Fig.12. Boiled and sterilized wheat grains

Fig.13. spawn of mushroom

4. Substrate Preparation and Spawning

Substrate Preparation differs from Mushroom to Mushroom

Oyster Mushrooms

There are several edible Pleurotus spp. viz. *Pleurotus ostreatus, Pleurotus flabellatus, Pleurotus florida, Pleurotus platypus* etc. which are well known for their delicacy and flavour. These species are being grown in commercial scale in various countries. However, in India, *Pleurotus florida* and *Pleurotus platypus* are most popular for commercial cultivation.

Substrate Preparation

Oyster mushroom can be grown on various substrates viz. wheat/ paddy straw, maize stalks, maize cobs, cotton waste, wooden logs, sawdust and vegetable plant residues. Since paddy straw is easily available throughout the year in most parts of the country, it is widely used. One should always prefer fresh and well dried paddy straw.

Soaking

Paddy straw chopped into 3-5 cm. Pieces is soaked in fresh water for 8-24 hours. Old, broken and rotten straw and stagnant water should never be used. Wet substrate is spread on a wire mesh to drain off excess moisture.

Pasteurization

Water is boiled in a wide mouth container such as tub or drum. The wet substrate is filled in gunny bag or basket and closed. The filled bag or basket is dipped in hot water (80-85°C) for about 30-60 minutes and to avoid floating it is pressed with the help of a wooden piece. After pasteurization, excess of hot water should be drained off in a container so that it can be used for other sets. Care should be taken to maintain boiling water temperature, 80-85°C, for all sets to achieve pasteurization.

Pasteurizes substrate is kept inside the chamber where bag filling and spawning are to be done. Once it cools down to room temperature it is filled in bags. The moisture content should be 60-70%.

Spawning

When pasteurized substrate has cooled down to room temperature, it is ready for filling and spawning. Polythene bags (35x 50 cm, 100-150 gauge) or polythene bags (35 x 50 cm, 100 gauges) may be used for its cultivation. Two percent spawn on the basis of wet straw i.e. one 500 ml, bottle spawn (200-250gm) for 10-12kg wet straw is used. Spawning can be done as follows.

Surface spawning

Bags filled with substrate are opened and 2% spawn is broadcasted on top and a little ruffling is done to mix it on the top layer (2-5cm thick) and closed.

Layer spawning

Substrate is filled and gently pressed to a depth of 8-10cm. and spawn is broadcasted above it. Similarly, second and third layers are put and simultaneously spawned and bags are closed. This is more suitable when pasteurized straw is filled in bags.

Thorough spawning

Pasteurized straw is mixed with 2% spawn and filled in bags. It is gently pressed and closed for spawn running.

Spawn bags should be stacked in racks in dark, neat and clean place in close position. Temperature of 25±2°C and relative humidity of 80-85% should be maintained by spraying water on walls and floor. At this stage fresh air requirement is minimum. It takes about 20 days when bags will be fully covered with white mycelium.

Cropping, Picking and Packing

Once bags are fully covered with mycelium, they are transferred to cropping room and polythene covers are removed. The open blocks are kept in racks about 20cm. apart. The racks should be 60cm. wide with a gap of 50-60cm. between two shelves. There should be 60-75cm.gap between two rows of racks for working. It grows in a temperature range of 20-33°C (optimum 25±2°C) relative humidity of 80-85% is maintained by spraying water twice a day, but watering on blocks should be avoided for the 2-3 days. After 2-3 days, light mist spray of water is given on the blocks. This is time when small pinheads appear. Once pinheads are 2-3cm. size a little heavier watering is done on blocks and further watering is stopped to allow them to grow. This helps to avoid bacterial rotting. However, the optimum relative humidity must be maintained inside the cropping room. Once mushrooms are 6-8 cm in size, they are plucked. Mushrooms should not be allowed to shed spores as it results in poor quality of mushrooms. After harvesting the first flush about 0.5-1 cm. outer layer is scraped. This helps to clean remnants of first crop and to initiate second flush. Second flush will appear in about 10 days of the first flush, similarly the third and fourth flushes will appear in 8-10 days' interval.

Mushrooms should be harvested before watering and their lower portion cleaned with a dry cloth. Mushrooms are packed in perforated (5-6 small holes) polythene or polypropylene bags. It should be sent to the market while fresh. This variety can also be sun dried by keeping fresh mushrooms in the sun for 2 days. Even polythene sheets may be spread at about 30-40 cm. above the mushrooms for quicker dehydration in the sun. It can be mechanically dried in the drier at 50-60°C. The dried products can be packed in polythene bags for marketing. Dried mushrooms should be soaked in water for 30 minutes before use.

Fig. 14. Fully grown Oyster mushroom

Paddy Straw Mushroom Cultivation

Paddy straw mushroom (*Volvariella spp.*) is grown in tropical and sub-tropical Asian countries viz. China, Korea, Honkong, Philippines, Malaysia, Singapore,

Thailand, Burma and India. Its cultivation is primitive and cultivation technique is still less advanced. There are two species commercially cultivated i.e.*Volvariella volvacea* and *Volvariella diplasia*. In India, it can be cultivated during summer and rainy seasons in North and throughout the year in the South, wherever favourable temperatures (25-38°C) are available.

Substrate Preparation

Paddy straw mushroom originally grew on paddy straw so it is called "paddy straw mushroom" or "straw mushroom". It can also be grown on various substrate other than paddy straw viz. Combination of paddy straw and cotton waste compost, sugar begasse, dried banana leaves, water hyacinth etc. Since paddy straw is easily available and performs well, it is extensively used. Fresh and well dried straw should be preferred as half dried or decaying straw may harbour moulds which may affect the crop.

Soaking

Paddy straw bundles (weighing 1 kg) are prepared using whole straw and tied at two places to make it firm. These bundles are soaked in fresh water for 8-16 hours. Some heavy material like stone or logs should be kept above the bundles to avoid floating. Wet bundles are kept on the cement floor or on raised supports to drain off excess water.

Pasteurization

As mentioned earlier, pasteurization helps to eliminate harmful microbes, nematodes and insect pests and helps in getting higher and almost constant yield. Substrate can be pasteurized either by dipping in hot water or treating with chemicals. Water should be boiled in a wide mouth container and wet bundles should be dipped in hot water(80-85°C) for 30-60 minutes to achieve pasteurization. In chemicals pasteurization, bundles are treated with formaldehyde solution (5%) 24 hours prior to soaking. Bundles are kept on a raised support to drain off excess water and once cooled to room temperature, bed can be prepared. Cotton seed husk or chicken manure (5%) is used as supplement and should also be pasteurized by dipping in hot water (80-85°C) for 30-60 minutes.

Bed Preparation and Spawning

In the cultivation of paddy straw mushroom, bed preparation and spawning is done simultaneously. Four to six bundles are kept on 75-90cm. wide platforms lengthwise with butts on one side. As soon as one layer is completed supplement (5%) and spawn are broad casted. For spawn preparation either chopped paddy straw or wheat/ sorghum grain is used. Straw spawn is sprinkled over with a small quantity of gram dal powder. A second layer is made by placing the bundles with the butt ends on opposite sides. Again spawn is broadcasted and a third layer of bundles is placed at right angles to the previous layer i.e. crisscross fashion, and the fourth layer is placed n the third layer with the opposite butt ends. It is also spawned in the same way.

Another layer of bundles is placed with the butt end at right angles to the previous one. This layer is spawned all over and a thin layer of straw is placed to cover it. Finally, the bed is pressed down. As much as 350-400g of grain spawn is required for spawning one bed. The bed is then covered with a polythene sheet. Watering depends upon the humidity of the air. Mostly no watering is required for first 3-4 days. Watering of the bed and the temperature within the beds are very important. Bed temperature should be 30-35°C. Small buttons start appearing 7-10 days after spawning. At that time polythene sheet is removed from the bed. They remain in the bottom for 4-5 days then grow into full size. When the volva is about to rapture it is picked by gently twisting the fruiting body.

Fig. 15. Volvariella

https://upload.wikimedia.org/wikipedia/commons/0/0c/Volvariella_sp._%2843811069950%29.jpg

White Button Mushroom Cultivation

The white button mushroom is more acceptable to the consumer and fetches a higher price. For its successful cultivation, careful attention must be paid to four main aspects, namely- good compost, pure and productive spawn, proper environmental conditions such as temperature, relative humidity and fresh air, and hygiene of the farm. Indifference or neglect of any of these may result in failure of the crop. The details of the different stages of its cultivation are given below

Compost Preparation

The substrate in which mushroom mycelium grows is a specially prepared material called compost. The purpose of composting is to convert various raw materials into a selective, homogenous, both physically and chemically suitable substrate for mushroom mycelium. Some compost formulae based on paddy straw/ wheat straw, horse dung manure are given below,

Formula 1:

Wheat straw/paddy straw	1000kg
Or	
Paddy straw + dried maize stalk	500 + 500 kg
Ammonium sulphate or Calcium	30kg
Ammonium nitrate	
Super phosphate	10kg
Urea	15kg
Wheat bran	100kg
Or	
Rice bran	165 kg
Gypsum (Ca SO_4)	40kg
Calcium carbonate (chalk powder)	30kg

Note: when paddy straw is used, add 20kg cotton seed meal or 40 kg rice bran or 40 kg neem cake.

Formula 2:

Paddy straw	1000kg
Rice bran	166kg
Urea	16.6kg
Neem cake	40kg
Gypsum (Ca SO_4)	40 kg

Formula 3:

Paddy/ wheat straw	1000kg
Chicken manure	750kg
Wheat bran	40kg
Gypsum (Ca SO_4)	30 kg

Formula 4:

Strawy horse dung(straw 25-30%)	1000kg
Urea	6.8kg
Rice bran	20kg
Gypsum (Ca SO_4)	25 kg

There are two popular methods of compost preparation

Long method:

It takes about 26-30days and 7-8 turnings are given at varying intervals without pasteurization which may be easily infested by weed moulds, pathogens, insects, mites and nematodes. This method is popular with small growers who cannot afford boilers for pasteurization. The schedules of turning are given below.

Day-2: the straw/ manure is spread and watered in layers making a stack of 1-meter height. Next day, day-1, it is turned and again watered to saturate it.

Day 0: at this moisture content of straw should be 75-77%. Mixtures of fertilizers and bran is mixed with straw thoroughly or in layers and stacked in a dimension of 1.5-1.8m (width) x 1.5-1.8m (height) and length may depend on quantity used. Dry spots are sprinkled with water. Stack is gently pressed to have firm pile.

Day+4: first turning. A portion of stack should be broken, thoroughly shaken and mixed with water if required and stacked in layers. Subsequently, remaining portions should be stacked to have stack of 1.5-1.9 m x 1.5-1.8m x length.

Day+7:second turning is given as above and gypsum is added in layers. Watering is done if required. Day +10, third; day +13, fourth, day +16, fifth; day +19, sixth; day +22, seventh; day +25, eight turnings are given. By this time ammonia smell disappear. However, in case ammonia smell is noticed it should be further turned for one or two turnings to get free from ammonia. At this stage, compost will be dark brown in colour, with moisture content about 65-70%.

Short method:

This is the most popular method for commercial production of white button mushrooms. It is completed in two phases.

Phase I:

Day-4: straw/ manure (horse dung/ chicken manure) are spread n layers and thoroughly watered to make a make a stack of 1m height. Excess water should be stored in a pit in a corner of composting platform and reused to water.

Day-2: stack is turned and watered thoroughly and the seepage water collected should also be used.

Dayo: at this stage the moisture content of substrate (straw/ manure) should be 75-77%. Fertilizer, bran is mixed in layers and watering is done if required and standard stack (1.5-1.8 x 1.5-1.8 x length) is prepared.

Day+4: first turning

Day+7: second turning, gypsum is added. Watering is done if required. Stack is gently pressed to maintain loose pile.

Day +10: third turning. Stack is broken and thoroughly shaken and filled in trays for phase II. At this stage compost will have a strong smell of ammonia with a moisture content of 70-72% and is light to dark brown in colour. In another schedule, phase I is extended to two more turnings i.e. day +13 and day +16 when ammonia smell will be moderate. Compost is filled for short phase II (3-4 days).

Phase II

Phase II helps to complete fermentation of the compost and to eliminate harmful microbes, insects and pests. Compost filled trays are stacked in the room maintaining75cm. gap on the side blower and steam outlet is provided and 45 cm. on its opposite side with ventilators at ground level and top on the narrow side fitted with mesh and shutter. Steam is released and blower is switched on to circulate inside air and 10% fresh air is introduced through air circulating duct. Air temperature is maintained at 45±2^{0}C for 2-3 hours to stimulate compost fermentation. Next day all ventilators are closed and air temperature is raised to 58^{0}C which help set compost temperature to 60^{0}C. It is maintained for 2-4 hours. Temperature is monitored from outside. On completion of this stage (pasteurization) ventilators are opened and again fresh air (10%) is introduced to reduce temperature slowly from 55-50^{0}C. this stage is most important to achieve fermentation in another 2-3 days, when compost becomes free ammonia. One may notice sweet smell of thermophillic microflora. At this stage compost should be dark brown in colour, free from moisture content 68-70%. Blower is switched off and compost is allowed to cool down below 28^{0}C in a day or two.

Spawning and Spawn Running:

Spawning means putting spawn i.e. mushroom mycelium coated on sterilized substrate/ grain in the compost. Three methods of spawning have been employed.

- **Surface spawning:** Upper portion of compost is lifted to a depth of 2.5-5cm. and spawn is spread and ruffled with fingers into 2.5-5cm. of compost and again covered with lifted compost and pressed to have uniform top surface.
- **Layer spawning:** In this method spawning is done in two layers. First layer is spawned at about 8-10cm. down from the top and the next layer is spawned as done in surface spawning.
- **Thorough spawning:** This is done by mixing spawn evenly and throughout the compost.

After spawning compost is covered with newspapers and water is sprayed to keep the papers moist. A temperature of 25±2°C and relative humidity of 85-90% is maintained by spraying water on trays and walls. During 24-48 hours after spawning cultivator is careful about compost temperature, if temperature goes above 30°C water is sprayed and ventilators are opened as temperature more than 34°C is injurious to mushroom mycelium. Twice in a week compost surface is checked if it is dry paper is removed and water is sprayed with mist sprayer. Mushroom mycelium starts spreading from the spawn and permeates through the compost. It is called "spawn running". It covers the entire compost as whitish thread-like structures with a fine fluffy appearance of the mycelium in 14-18 days. At this stage cover is removed and casing is done.

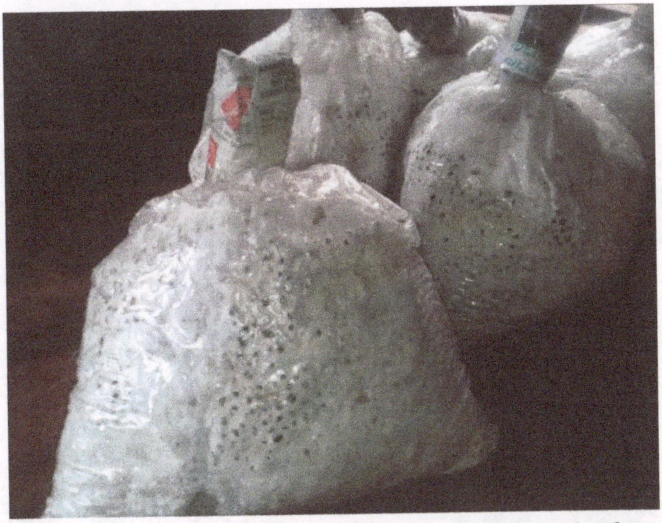

Fig. 16. Polythene bags showing spawn ran and growth of mycelia

Casing:

The term casing means covering the compost covered with mushroom mycelium, with a layer (4cm.) of soil or other materials such as old (2 years) spent compost, cow dung, coir dust, farmyard manure etc. which retains moisture and helps in gaseous exchange. Lime or chalk power is added to maintain a pH of 7.5-8. Casing material is chemically pasteurized by drenching casing material with formaldehyde solution (5%) two weeks before casing so that at the time of casing it is free from formalin. However, steam pasteurization (65°C for 5-6 hours) is most effective. Casing material slightly moist (about 20%) is filled in trays inside a small chamber and steam is released to maintain casing material's temperature at 65°C for 5-6 hours. On cooling casing material is spread in about 4-5cm. thickness on compost after removing the newspaper. The temperature of 23±2°C and relative humidity of 85% is maintained and light watering is done to keep casing material moist. Over watering is avoided.

Post composting supplementation of the substrate at spawning and casing is done with cotton seed meal, soybean meal, lucern meal, corn gluten meal, cracked soybean, de-oiled soybean, groundnut cake, cotton seed cake, inorganic fertilizer, yeast extract, sodium chloride, ammonium sulphate, vitamin B complex, feather meal, bone meal, etc to increase yield.

Cropping:

It takes about 10 days when mycelium covers casing layer and this is the time for pinning and cropping. Temperature is lowered to 16±2°C and fresh air of compost is introduced. Relative humidity of 80-83% is maintained. This results in pinning. During this period light watering is done. When pins are pea size little heavy mist water spraying should be done. Once mushrooms are pea size they take another 3-5 days to attain 2.5-4 cm. when they are harvested. Mushrooms appear in flushes of 3-5 days' duration each at intervals of 7- 10days.

Fig. 17. ***Agaricus bisporus***

http://images.google.co.in

5. Harvesting, Marketing and processing of Mushrooms

Generally harvesting is continued for 8-12 weeks. Many growers prefer shorter crops of 5-6 weeks as initial weeks produce major part of the crop. An average yield of 4-5kg /tray (0.9 x 0.6x 0.18M) is generally obtained in 8-10 weeks. Mushrooms are packed in perforated polythene bags and marked fresh or canned in brine solution.

6. Storage of Fresh Mushroom

The white button mushrooms are covered with PVC film and stored at 15 to 21°C. This increases the life of mushroom from 5 to 7 days.

If the mushrooms are packed in 0.02 to 0.03 mm densepolythene bags with nitrogen and stored at 0°C, its self life can be increased up to 5 weeks. As the temperature rises, self life span of mushroom starts to decrease. The self- life of mushroom with nitrogen is reported to be 2 weeks at 15°C.

Caning

Canning process is followed for small sized white button mushrooms. The mushrooms selected for canning process, have 0.4 to 1 cm long stalks attached to the cap. After picking, the stalks of the mushrooms are removed by cutting close to the button and mushroom showing and blemishes are rejected. These are washed gently to remove soil particles sticking on the cap. These buttons are then blanched for 5 minutes. Blanching is done to discolouration which may cause shrinkage to the tune of 25-30% shrinkage is unavoidable. Blanching also improves the texture and gives greater drain weight in the can. White button mushroom is blanched for 5 minutes either in steam or boiling water and put in cold water to prevent blanching loss. Blanching may be done with following treatments

- Immersing mushroom in 0.1 to 0.2% citric acid solutions for about 2 minutes and cooled immediately in cold water.
- Immersing of mushroom in brine solution containing ascorbic acid and EDTA @ 1% and 1000mm, respectively.

Dehydration

The fully mature mushrooms with stalks are used for dehydration. They are gently washed and dried either in sun or in a mechanical drier. The dehydration is done at 50-60°C. The dried mushrooms should not contain more than 5% moisture. The mushrooms may be blanched before dehydration. The dipping of mushrooms for 5 minutes in 400 ppm chlorine aqueous solution and 300 ppm Sulphideaqueous solution after blanching is reported to maintain the white colour and negligible bacterial population. After dipping in these solutions mushrooms are dried. Weight of dried mushrooms is reduced to about 1/8 to 1/12th of theiroriginal fresh weight, depending on the species of mushrooms. The dried mushrooms are packed in hermetically sealed air tight tins for quality retention and then stored at a cool dry place. The dehydrated mushrooms retain their original colour andflavour for six to seven months during storage. The dried mushrooms can also be powered by grinding and used in making mushroom soups. The dried mushrooms are reconstituted by immersing in sucrose solution with ascorbic acid.

Freezing drying

Washed button mushrooms (*Agaricus bisporous*) are dipped in .05 % to sodium metabisulphate and 2% sodium chloride solution for 30 minutes.

Freezing: harvestedmushrooms are cooled to 2 to 4°C, washed with aqueous solution of 50 ppm chlorine then blanched in 0.1% soln of citric acid for 3 to 4 minutes, and then they are frozen at -34°C for in minutes by dipping in Freon-12 (quick freezing).

Stepping preservation: harvested mushrooms are washed and blanched in hot water containing 0.1% citric acid for 5 minutes, washed in cold water and canned in 0.05% citric acid and 100 ppm SO_2.

Factors affecting mushroom cultivation

Temperature

Table 8. Temperature and substrate requirement

Mushrooms	*Temperature*	*Substrate*
Agaricus brunnescens	25°C (V), 16-18°C (F)	Compost
Pleurotus florida	25°C (V),20-22°C (F)	Raw straw (wet)
Pleurotus ostreatus	25°C (V),8-15°C (F)	Raw straw
Lentinus edodes	25°C (V),16-19°C (F)	Logs
Flammulina velutipes	25°C (V),10-12°C (F)	sawdust
Auricularia spp.	25°C (V),18-25°C (F)	Logs and sawdust

White button mushroom- optimum temp: 22-25°C, during mycelium growth-28°C, above 34°C is lethal

Oyster mushroom- 23-28°C, temp, above 34°C is lethal

Paddy straw mushroom- 30°C

Relative humidity

- During spawning- 90-95 %
- During cropping more than 80-90 % causes bacterial blotch

Moisture content of the substrate- 68-70%

Ventilation

- Good spawn- single specimen
- Good substrate-Paddy straw, sugarcane trash, maize straw, banana leaves.

Common Diseases of Mushrooms

Mushrooms are subjected to many fungal, bacterial and viral diseases. Even a little carelessness in different steps during cultivation can play havoc with the mushroom. Improper pasteurization of compost and casing soil can be the major source of diseases. Once the disease is introduced in the farm the secondary infection can be carried out by different agencies, such as air, water, machines and workers. The important fungal, bacterial and viral diseases are as follow.

Table 9. Diseases of mushrooms

Diseases	*Causal organisms*
Wet bubble	*Mycogone perniciosa*
Green mold	*Trichoderma viride*
Mildew of cobwel disease	*Dictylium dendroides*
Bacterial pit	*Pseudomonas fluoredcens*
Bacterial blotch	*Pseudomonas tolaassi*
La france disease Die back, watery stipe	Viral disease
Insect pests	Sciarids, phorids, Spring tails, cecid, mites, nematodes

Chapter 5

Economic Importance of Mushrooms

Nutrient Content

Mushrooms are the natural God given health food to mankind. At present mushroom has become an important menu in human diet. Edible mushrooms are valuable sources of nutrients and are recognized as source of protein rich food. It also provides carbohydrates, salts and vitamins. Romans considered mushrooms as "Food of God" and as such, they served mushroom preparation to their guests on important occasions. It is also called "Vegetable meat". The protein content is 3.7g/100g of edible portion, and much higher as compared to various vegetables. Mushrooms proteins have a digestibility of 60-70%. In general, 100g of fresh mushroom contains the following: Carbohydrate 1-2g, Proteins 5-8g, Fat 0.2-1g, Vitamins 5-15g (Thiamine, Riboflavin, Nicotinic acid, Folic acid), Minerals 0.5- 1g (Na, K, P, Ca).

Because of the low caloric values (18-20 kcal/100g), little starch and negligible amount of fat, mushroom is ideal diet for diabetics. The abundant quality of folic acid is considered vital in treating anemia. "High amount of fibre in rich food is helpful for slimming as food gives a feeling of fullness. Fibre rich food also increases general health and fitness". Edible mushrooms contain high level of dietary fibre, substantial amount of protein, vitamins and minerals but are low in fat. They also have various properties for health benefits such as anti-oxidative, anti-tumour and hypercholesterolic effects.

Some of the Mushrooms Preparations

Table 10. Some mushroom preparations

Mushroom Omelets	Mushrooms And Peas
Stuffed Potatoes	Mushrooms And Paneer
Mushrooms On Toast	Mushroom Samosa
Mushroom Soup	Liver With Mushroom
Mushroom Pakoras	Mushroom Manchurian

Mushroom Pulao	Mushroom Mutter Paneer
Mushroom Pickle 1	Mushroom Masala
Mushroom Pickle 2	Matar Mushroom
Cheese Mushroom Biryani	Mushroom Korma
Mushroom Broccoli Vegetable	Mushroom on Toast
Mushroom Rice	Mushroom Pakoras
Mushroom Curry	Capsicum Mushroom Fry
Mushroom Paneer Gravy	Mushroom Rolls
Fried Mushroom	Mushroom Baigan Bharta
Malai Mushroom Curry	Mushroom Sandwich
Mushroom Cashew	Walnut Patties

Mushroom Toxins

But ever since humans have become curious about the beautiful mushrooms, poisoning by consuming them has become quite a phenomenon. By trial and error method, humans have learnt to distinguish the edible mushrooms from the poisonous ones. There is no clear cut easy method to identify edibility of mushrooms. According to Gogoi *et al.*, and others a poisonous mushroom will have a sac or cup like structure called volva at the base of the stalk, it will also have skirt like ring on the stalk near the cap, Gills will be white in colour and it will grow singly or in twos and edible mushrooms will have either ring or volva or none of these.

Other features of poisonous mushrooms are, they comparatively soft, colourful and very attractive, exudation of milky substances from the damage fruit bodies, bitter or sour in taste and have unpleasant smell and generally insects, ants and other animals do not feed on them. Some traditional methods to distinguish edible and poisonous mushrooms are like, 1. If brinjal pieces or flowers are dipped into mushrooms soaked water, if it turns blackish then mushroom is poisonous. 2. If poisonous mushroom is soaked in salt water, water turns blackish colour. 3. Silver or copper particles will turn blackish when immersed in the dish of poisonous mushrooms. 4. Milk curdles immediately in contact with poisonous mushrooms. But these criteria are said to be not very reliable therefore care should be taken while collecting mushrooms from the field and no one should attempt to eat a wild mushroom without being 100% sure of its identity.

With the advancement of science, some of the poisonous mushrooms were analyzed for their chemical composition. This resulted in the discovery of many toxic substances, which are responsible for mushroom poisoning. Some of these toxins, in the pure form, are fatal even in the smallest dose while others impair the various physiological functions of our body. Some of these toxins have found use in both Homoeopathic and Allopathic systems of medicine, while others are employed in the biochemical research. Some of mushroom toxins, their effects and probable uses are briefly discussed below.

Mushroom poisoning by *Amanita phallodes, A. verna, A. virosa* and *Galerina spp.* is often deadly, mainly due to toxic substances like amatoxin and phallodin present in fresh fruiting bodies (basidiocarps). Amatoxin damages the nuclei of the liver cells, by inhibiting RNA Polymerase B, essential in protein synthesis. This property of amatoxin makes it an important tool in biochemical studies. Phallodin also damages the liver cells but is not toxic as amatoxin. Similar effect on liver cells and kidney is produced by *Gyromitrin,* which is produced by consuming raw or half-cooked Ascomycetous fungus *Gyromitra esculenta.*

Another potent nephrotoxin is orellanin, which damages the kidney permanently when basidiocarps of *Cortinarious orellanus* are consumed. Interestingly, *Coprinus commatus* will not go along with alcohol as it contains a toxin called coprine which inhibits alcohol dehydrogenase in the liver, affecting the conversion of alcohol to acetate. An increased level of aldehyde in blood causes alcohol antagonistic symptoms and can be used to treat alcoholics.

Some mushrooms when consumed produce hallucinogenic effects. *Amanita muscaria* (fly agaric) has toxins like muscarine, ibotenic acid. When the basidiocarps are dried secondary metabolites like muscimol and muscazone, very potent hallucinogens, are produced. All these toxins stimulate the aminobytric acid receptors of the cells of the central nervous systems. Muscarine is used in Homeopathy, to treat mental disorders. Both ibotenic acid and muscimol have insecticidal properties and can be used as insect repellants. This fungus is believed to have been used by Aryans, as a source of soma, a divine drink. Psychotropic fungi, like species of *Psilocybe*, contains chemical like Baeocystin, Psiocin and Psilocybin which are potent serotonin antagonists when taken even in very small doses. It gives the same effect as that of Lysergic acid. Therefore, it is often used in Psychotherapy and neurological research.

It is surprising to note that there are reports of edible mushrooms having some toxic substances as in the widely cultivated button mushroom (*Agaricusbisporus*). It has agaritine, a hydrozene derivative considered as a carcinogen. Paddy straw mushroom (*Volvariella volvacea*) contains a thermolabile audiotoxin and the mushroom enokitake (*Flammulina velutipes*) has flammotoxin. This calls for a reassessment of the wisdom of their use as food. But consuming shiitake mushroom (*Lentinus edodes*) is beneficial. This is due to lentinan, an antitumerous substance, and eritadenine, a hypolipidemic substance which reduces cholesterol in blood.

Medicinal Value

There are reports of medicinal uses of different mushrooms. *Termitomyces robustus* when eaten serves as blood Tonic. Termitomyces globulus and *Termitomyces clypeatus* R. Heim. are preferred by the Asantes, Akyems, Akwapims and are also believed to lower blood pressure in hypertensive patients. *Termitomyces* is affluent source of sugar, protein fibre, lipid, and vitamin, mineral in addition to medicinal value which is used in lower blood pressure, rheumatism, kwashiorkor, obesity, and diarrhea and as purgative. The spores of *Lycoperdon perlatum* Pers.: Pers. and *L. pyriforme* Schaeff.: Pers. are used as powder to heal wounds and to cure baby rashes. Several species of Lycoperdon are used to stop bleeding in a fresh wound. *Termitomyces* species is an important source of enzymes of industrial importance such as xylanase, amylase, and cellulase, antioxidant compounds such as polyphenol and vitamin C, protein 31.4–36.4% and immune stimulatory agent. Indigenous peoples are utilizing mushroom for the treatment of different type of diseases and also as an aphrodisiac and tonic. Following table shows medicinal uses of different mushrooms by different ethnic groups of Southern Ghana.

Medicinal uses of different Mushrooms used by Ethnic Groups

Table 11. Medicinal uses of mushrooms

Name of the Mushroom	*Ethnic group*	*Medicinal use to treat/ for*
V. volvaceae	All tribes	lowering high blood pressure, Blood tonic
Termitomyces globulus	Asantes, Adangbes Akyems, Akwapims	Blood tonic, kwashiorkor, Lowering high blood pressure
T. microcarpus	Asantes	Blood tonic
T. clypeatus	Sefwis	Rheumatism, diarrhoea, Lowering high blood pressure
T. robustus	All tribes	Blood tonic
Coprinus disseminatuss/ Schizophyllum commune	Asantes, Brongs	Blood tonic, purgative, eye Infection, pregnant women
Collybia sp.	Brongs, Asantes	Blood tonic, fever
Pleurotus tuber-regium	Brongs	Asthma, fattening Underweight children, lowering high blood pressure
Auricularia auricula-judae	Brongs	Blood tonic

Macrolepiota procera (Scop.) Singer has been reported to demonstrate anti-tumor activity to human body and exhibit an antibiotic activity against gram negative bacteria, the fruiting bodies have been extensively used for manufacturing traditional foods and medicines. *Macrolepiota procera*(Scop.) Singer. is used for edible purposes, the fruiting body of it contains proteins, iron, zinc, chitin, chitosan, fiber, vitamins and minerals to support the health and sustain physiological homeostasis of human body.

Mushroom role in Bioremediation

Mushrooms are very specific to their nutrient source thus they grow on particular habitat for their nutrient requirement. They grow saprophytically on various dead substrates and parasitically or symbiotically on or with other organisms. Their fruiting bodies occur either above the substratum or hidden under the substratum.

Mushrooms can be classified into three main groups based on the natural habitat. Humicolous (humus inhabiting), Lignicolous (wood inhabiting) and Coprophilous (dung inhabiting). The natural habitat of mushrooms is helpful to know the nutrient requirement of a particular group of mushrooms and thus their chemical composition also can be identified.

Without fungi, we would not be able to go for a stroll in the bush. Mushrooms are the recyclers in nature, they break down wood and humus back into their original components and therefore not only making it easier for our little adventures in the wild but they also provide food for living plants by returning dead trees and forest littler to simple organic materials in a form suitable for plant use. They take turns on the same tree, because they differ in their ability to break down certain materials. So, when a tree falls down, it may be attacked first by, say, an Oyster Mushroom, which will slowly decompose the tree until it reaches a stage of decomposition that is not longer suitable for it to sustain its existence. This succession continues until there is no more wood left.

Chapter 6

Mushrooms of Jharkhand

Rainy season in Jharkhand normally starts from last week of June. As soon as the rainy season starts varieties of mushrooms appear in the forest. Villagers are well familiar with this phenomenon; they go to the forest and look for these wild edible mushrooms. *Lycoperdon, Astraeus, Boletus edulis* Bull., *Macrolepiota procera* (Scop.) Singer, *Russulasanguinea* (Bull.) Fr. etc are found with first rain itself. *Termitomyces* and later variety of *Astraeus*appear a bit later. Mushrooms are found from July to September. It was found that *Astraeus, Termitomyces heimii* K. Natarajan *and Termitomyces albuminosa* (Berk.) R.,were sold abundantly in the village market with a high price.

The price of 1 kg of *Astraeus*and Termitomyces were 320-400 rupees in the first week of July in 2011. According to the villagers *Termitomyces clypeatus* R. Heim is the most delicious mushrooms but it is not found abundantly in the forest. During the field study many edible as well as poisonous mushrooms were also encountered. These mushrooms have local names in local languages.

Table 12. The scientific and common names of mushrooms in *Sadri*, *Mundari*, *Khurukh* and *Kharia* languages

Scientific name	*Sadri*	*Mundari*	*Khurukh*	*Kharia*
Astraeus odoratus (early variety)	Rugra, Putu, Jait Putu	Jatiputukoi	Mani	Jait Pudub
Astraeus hygrometicus (late variety)	Rugra, Putu, Kherputu	Pundiputukoi	Chirya	Rabtob
Termitomyces clypeatus R. Heim.	Nabhi khukhri	Lechreu:d	Kudaosa	Chirko
Termitomyces heimii K. Natarajan	Patiyari	Pidiu:d	Chelariosa	Konterlu
Termitomyces albuminosa (Berk.) R.	Barka Khukhri	Indiu:d	Barkaosa	Mahau:d
Lycoperdon sp.	Tumba putu	Tumbachokerote	Tumbaosa	Tumbapudub
Boletus edulis Bull.	Jamun Khukhri	Kudau:d	Jambuosa	Kudau:d

Scientific name	*Sadri*	*Mundari*	*Khurukh*	*Kharia*
Macrolepiota procera (Scop.) Singer		Bushuu:d		
Russula sanguinea (Bull.) Fr.	Patra	Patrau:d, Simdali	Bhadraosa	Patra, Kokrochundi
Armillaria mellea (Vahl) P. Kumm	Kulthi Khukhri	Horezu:d	Arsgaosa	Koraizu:d
T. microcarpous (Var).	Bala Khukhri	Gitilu:d	Baluosa	Torezu:d

Some of the mushrooms collected are as follows. *Macrolepiota procera* (Scop.) Singer. It has tall stipe (up to 25 cm) and a large cap with rusty brown coloured umbo. It is coated with pink granules; spore print is white. June is the best month for this mushroom. Rain always precedes the emergence of this mushroom. Maximum and minimum temperatures requirement for fruiting are 30°C and 24°C respectively. It tastes good but not much desired for consumption. It grows on the humus rich soil.

Its taxonomic position

Class: Basidiomycetes

Order: Agaricales

(Hymenomycetes)

Family: Agaricaceae

Genus: Macrolepiota procera (Scop.) Singer.

The *Termitomyces* is an important genus of fleshy fungi and it is most abundantly found in Jharkhand. *Termitomyces* species is one of the choice edible mushrooms which grow wildly in symbiotic association of termite. Mainly four species of edible *Termitomyces* are found in Jharkhand. There is great difference in the morphological characters among four species. Three of them are described below.

Termitomyces clypeatus R. Heim., The distinguishing character of this mushroom is the prominent, cone - like pointed, dark colour projection at the centre of the cap. It is found between June and August. It grows from termite mounds. This is one of the most delicious mushrooms found in nature. This mushrooms is not found abundantly, mostly it is found singly. When it is fully grown it opens like umbrella. There is another mushroom which resembles this and looks like *Termitomyces fuliginous* which is poisonous therefore care should be taken when it is collected from the field.

In Jharkhand, this mushroom is known as *bada khukhri* (big mushroom). It comes up exclusively from termite's nests. A patch of veil is present at the top of the pileus. The skin of the cap peels off. The stalk is about 25 cm long and swollen near the attachment with the cap. The spore print is deep pink. It is a very delicious mushroom. It appears one month later after *Termitomyces heimii* K. Natarajan appears and this is bigger in size. Usually it appears when *Karam* festival is celebrated in Jharkhand in the month of September and this mushroom is very specific to this month. Its occurrence lasts for about 2-3 weeks only. It takes almost three days to grow into a full fledge fruiting body.

It looks similar to *Termitomyces albuminosa* (Berk.) R.but little smaller in size and occur at least one month before *Termitomyces albuminosa* (Berk.) R. appears. It has annulus and long stipe and locally known as *patiyari khukhri*. In Jharkhand, this mushroom is found most abundantly.

Table 13. Comparative study of three species of *Termitomyces*

***Termitomyces clypeatus* R. Heim,**	***Termitomyces heimii* K. Natarajan**	***Termitomyces albuminosa* (Berk.) R.**
• *Strongly spiniform perforatorium* long, central, solid, white to dirty brown fleshy and fibrous stipewhich may 3-7 cm long produced fruit bodies having 3-7.00cm long stipes (5.5-8 cm)	• *Perforatorium with umbo*	• *Sharply differentiated perforatorium* • white, solid stipe with up to 9 cm long and 3.2-4.5 cm in diameter • produces fruit bodies with significantly longer stipe (9.00cm) (6.00 cm)
• *Termitomyces clypeatus* R. Heim.produced fruit bodies with silky surface of pileus and irregularly lobed margin	• Stipe of *Termitomyces heimii K.* Natarajan is cylindrical creamy stuffed with thick annulus, up to 5 cm long and up to 1.5 cm wide	• *Termitomyces albuminosa* (Berk.) R. produced fruit bodies having entire margin with smooth pileus
• *Termitomyces clypeatus* R. Heim.produce orange grey to grayish orange colour fruit bodies	• Termitomyces heimii K. Natarajan (5.00 cm)	• *Termitomyces albuminosa* (Berk.) R. generated whitish grey, silvery shining fruit bodies
• *Termitomyces clypeatus* R. Heim. are free, white, pink at maturity, edge entire	• Pileus length up to 8 cm	• *Termitomyces albuminosa* (Berk.) R. produced crowded gills, free, white in colour
• Flesh of *Termitomyces clypeatus* R. Heim. are soft and inflated, hymenophoral trama regular	• *Termitomyces heimii* K. Natarajan produced fruit bodies with incurved margin and smooth type of pileus	• *Termitomyces albuminosa* (Berk.) R. generated white flesh, soft, hymenophoral trama bilateral while young, regular at maturity

***Termitomyces clypeatus* R. Heim,**	***Termitomyces heimii* K. Natarajan**	***Termitomyces albuminosa* (Berk.) R.**
• *Termitomyces clypeatus* R. Heim.shows pink spore	• *Termitomyces heimii* K. Natarajan produced cream colour fruit bodies	• *Termitomyces albuminosa* (Berk.) R.having pink spore print
• *Termitomyces clypeatus* R. Heim.lacking both an annulus and other velar remnants	• *Termitomyces heimii* K. Natarajan *produced* crowded, free whitish, turn pinkish with age, margin; serrulate	• *Termitomyces albuminosa* (Berk.) R. produced persistent annulus
• *Termitomyces clypeatus* R. Heim.having cylindrical, slender, surface pale grey, smooth, glabrous, solid, fibrous and up to 13 cm long pseudorrhiza	• *Termitomyces heimii* K. Natarajan produced white flesh, hymenophoral trama regular with thin walled parallel hyphae	• *Termitomyces albuminosa* (Berk.) R. produced hallow pseudorrhiza with blunt end
• *Termitomyces clypeatus* R. Heim.(11.53)	• Spores prints of *Termitomyces heimii* K. Natarajan are purplish grey • *Termitomyces heimii* K. Natarajanproduced thick and persistent annulus • *Termitomyces heimii* K. Natarajan produced significantly long and hallow pseudorrhiza up to 16 cm below ground levelhighest average weight of fruiting body was found in *Termitomyces heimii* K. Natarajan(11.60)	• *Termitomyces albuminosa* (Berk.) R. (11.20)

Taxonomic position

Kingdom: Fungi

Division: Basidiomycota

Class: Basidiomycetes

Subclass: Agaricomycetidae

Order: Agaricales

Family: Lyophyllaceae

Genus: Termitomyces

Lycoperdon is a genus of puffball mushrooms. Fruit body spherical or pear shaped with round top and tapering stem, 1-6 cm wide, 3-7 cm high. At maturity the powdery gleba, typically with light-coloured spores and a well-developed capillitium, is surrounded by a peridium consisting of from two to four layers. It grows in the beginning of the rainy season. Spores are formed in a basidium; basidia are borne within a container called peridium.

Lycoperdon grows fast and decays very soon. There seems to be no pore at the centre of the fruiting body for spore dispersal, fruiting body decays and the spores are released.

They should be consumed when they are very young and spores are not matured. *Lycoperdon pyriforme, L. Gemmatum and L. umbrinum* are common species.

Its taxonomic position,

Class: Basidiomycetes

Order: Lycoperdales (Gasteromycetes)

Family: Lycoperdaceae

Sub-family: Lycoperdae

Genus: Lycoperdon

It is commonly known as earthstars. The name comes from *geo* meaning earth and *aster* meaning star. The name refers to the behavior of the outer peridium. At maturity, the outer layer of the fruiting body splits into segments which turn outward creating a star-like pattern on the ground.

The inner peridium is a spore sack. In some species, the outer peridium splits from a middle layer, causing the spore sack to arch off the ground. If the outer peridium opens when wet and closes when dry, it is hygroscopic. It is found in the months of June and July. It is found abundantly in the *Sal* (*Shorea robusta* Gaertn.*)* forest. In Jharkhand local people call it *rugra or putu.*

There are mainly types of *Astraeus* in Jharkhand. One of them has less threat like mycelia and this appears as soon as rain starts in the month of June where as the other one which has got more net like mycelia appears after few weeks.

Taxonomic position

Class: Basidiomycetes

Order: Lycoperdales

(Gasteromycetes)

Family: Lycoperdaceae

Sub-family: Geastrae

Genus: Geastrum

Table 14. Comparative study of *Lycoperdon* and *Geastrum*

<table>
<tr><td rowspan="4">Mushroom</td><td rowspan="3">Basidiomycetes Spores in basidium Y</td><td>Fruiting layer exposed to air X</td><td rowspan="2">Stem like receptacle arising from a definite egg-stage, with volva at the base of the stipe X</td><td rowspan="2">z Fruit body opening at maturity with round holes at apex, plant star shape- Geastrum</td></tr>
<tr><td rowspan="3">Fruiting layer not exposed to the air Y</td></tr>
<tr><td rowspan="2">Stem like receptacle absent or rudimentary, volva absent Y</td><td rowspan="2">Fruit body opening at maturity with round holes at apex, plant not star shape, globular flask shaped, base sterile-Lycoperdon</td></tr>
<tr><td>Ascomycetes Spores in ascus X</td></tr>
</table>

Y= yes X= no

The edible *Boletus* of Jharkhand is always found in association with *Syzygium cuminii* Linn. Therefore, it is known as *Jamun Khukhri.* In size, this is the biggest mushroom of Jharkhand. It tastes good and appears in the beginning of rainy season. Spores are formed in basidium, basidia borne on hymenium. Its lining is present inside the tubes therefore they are called tube fungi. Tubes readily separate from the cap and from each other. The name is derived from the Latin term *Bōlētus* 'mushroom' from the Ancient Greek ultimately from *bōlos* 'lump' or 'clod'. The genus *Boletus* contains many members which are edible and tasty. This is a symbiotic Fungus.

Taxonomic position

Kingdom: Fungi

Division: Basidiomycota

Class: Agaricomycetes

Order: Boletales

Family: Boletaceae

Genus: Boletus

***Species: Boletus edulis* Bull.**

Russula sanguinea (Bull.) Fr. grows on the litter of the forest in the July and August months. People of Jharkhand call it *Patra* because it grows on the decaying leaves. This mushroom taste well when they are consumed when they are young but if they are eaten when they are fully matured they do not taste well. This is fast growing and decaying mushrooms. This mushroom is found in different colours like white, yellow and red colour. As some of the mushroom specialists say that if mushrooms are eaten by insects it is most likely that mushrooms edible. Some insects are found in association with them. Since these mushrooms are found in different colours their spawn too look colourful.

Taxonomic position.

Kingdom: Fungi

Division: Basidiomycota

Class: Agaricomycetes

Order: Russulales

Family: Russulaceae

Genus: Russula

Species: *Russula sanguinea* (Bull.) Fr.

Table15. Months and Temperatures when these mushrooms occurr abundantly

Name of the mushroom	*Months of occurrence (most abundantly)*	*Temperature*
Geastrum sp. (early variety)	June-July	25-30°C
Geastrum sp. (late variety)	July-August	25-30°C
Termitomyces clypeatus R. Heim.	July -August	25-30°C
Termitomyces heimii K. Natarajan	August- September	25-30°C
Termitomyces albuminosa (Berk.) R.	September	25-30°C
Lycoperdon sp.	June, July -August	25-30°C
Boletus edulis Bull.	July-August	25-30°C
Macrolepiota procera (Scop.) Singer	July-August	25-30°C
Russula sanguinea (Bull.) Fr.	July-August	25-30°C

Scope of Mushroom Cultivation in Jharkhand

Chapter 7

Jharkhand is a densely forest area with deciduous forest which prepares very rich humus and rainy season normally starts in the month of June or beginning of July therefore varieties of wild edible mushrooms are found in the forest in the beginning of the rainy season. Jharkhand is located 300 meters to 900 meters above the sea level and is thickly populated by different tribal groups. The mushroom flora of Jharkhand is very rich; the tribal people have very good knowledge of wild edible mushrooms. They gather plenty of wild edible mushrooms in the beginning of the rainy season and they sell and earn some money.

People over here are depended on agriculture. They cultivate paddy, ragi, pea, cowpea, maize, sorghum and French bean etc. thus they have plenty of substrate for mushroom cultivation. Temperature and weather condition is suitable for mushroom cultivation and man power is in abundance. There have been efforts to grow mushrooms and train local people in the institutions like Birsa Agricultural University, Kanke and Horticultural and Agro-Forestry Research Programme, Palandu and by many NGO groups and people have shown keen interest toward mushroom cultivation.

About the local mushrooms of Jharkhand very less research works have been done. As these mushrooms are very delicious and rich in nutrient, they might be used as food source and medicinal purpose.

The indigenous people over have rich knowledge of wild edible mushrooms. They collect them from the forest and bring them to the city market for sale but some people are hesitant to buy them especially when these mushrooms are collected from the field. The varieties of mushrooms are found here but they are seasonal and most of them are symbiotic mushrooms. They grow in association with some particular tree and with termites. They seem to be very much humidity and temperature sensitive. Therefore domestication of these mushrooms is not an easy job.

The study of the natural habitats shows that most of the wild edible mushrooms of Jharkhand grow on the soil enriched by the humus of the decaying leaves. It was observed that *Geastrum* abundantly occurs under the *Sal* tree and *Termitomyces* occurs under the trees where lots of decaying leaves are there. *Termitomyces* are observed growing on the termites' hill. *Lycopderdon*is found growing in the garden or uncultivated lands. *Boletus edulis* Bull.is found always in association with *Syzygium cuminii* Linn.

There are mainly two species of *Astraeus* and both of them are edible. One of them appears as soon as rainy season starts and has less mycelium, looks brownish in colour and grows singly whereas the other species which appears a bit later and has more mycelia, it is whitish in colour, grows in cluster. *Astraeus* is found abundantly in the *sal* forest. *Russula sanguinea* (Bull.) which is locally known as *Patra* (decaying leaves) is found in white, red and yellow colour. *Boletus edulis* Bull. is found always in association with *Syzygium cuminii* Linn. tree and this is the biggest mushroom of Jharkhand. Mainly four species of *Termitomyces* are found they are associated with termite's comb, they different from each other in many characteristics.

In Jharkhand there are institutions like Birsa Agricultural University, Kanke, Horticultural and Agro-Forestry Research Programme, Plandu, Ramakrishna Mission, Morabadi and SRI (Social Rural Industrialization), Bariatu, where mushrooms are grown and local people are trained to cultivate mushrooms. In the recent years, many NGO groups and private agencies have shown keen interest toward mushroom cultivation. Yet mushroom cultivation in Jharkhand is in its initial stage.

Fig. 18. Children collecting mushrooms from the forest

Fig.19. A person digging out a mushroom from the soil

Fig. 20. Women selling wild edible mushrooms at *Kachhari* in Ranchi

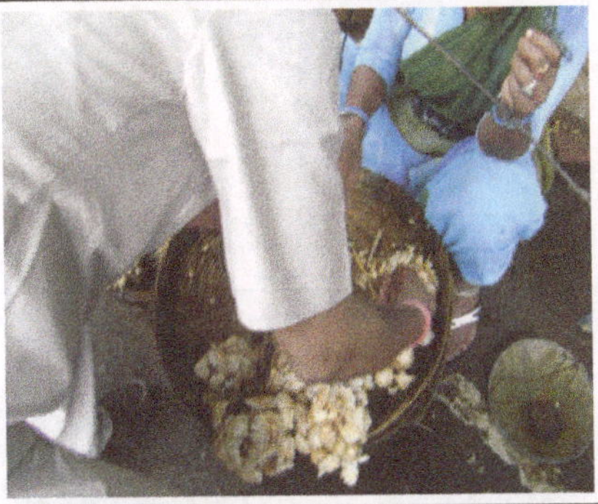

Fig. 21. Women selling wild edible mushrooms, *Rugra* and *Termitomyces* were being sold. Mushrooms were collected from Khunti and Pithoria areas.

Fig. 22.*Macrolepiota procera*

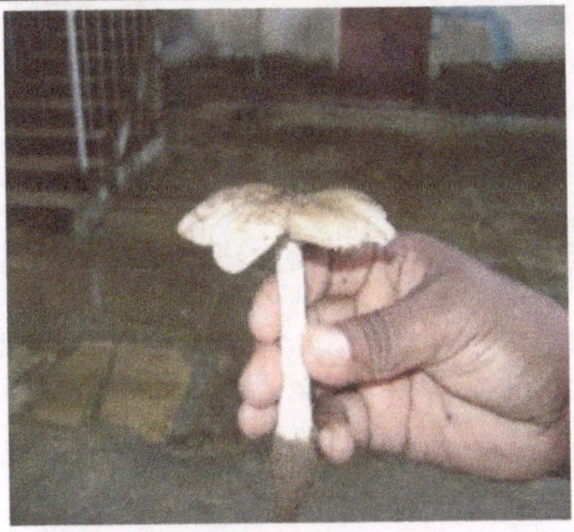

Fig. 23. *Termitomyces clypeatus* R. Heim

Fig. 24. *Termitomyces albuminosa* (Berk.) R.

Fig. 25. Stages in the growth of *Bada Khukhri*

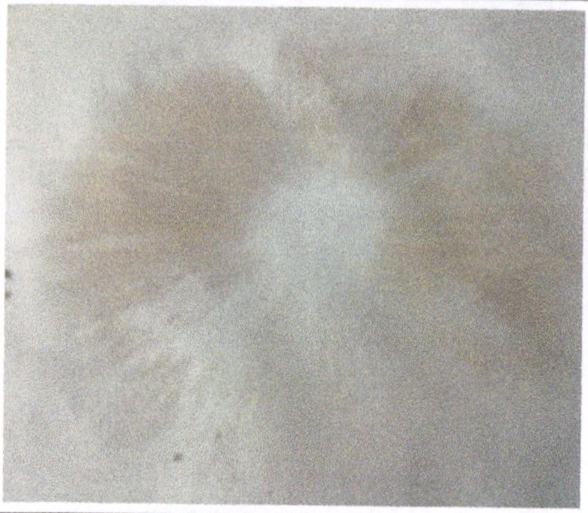

Fig. 26. Spore print

Fig. 27. *Termitomyces heimii* K. Natarajan

Fig.28. Growth of a mushroom on the culture medium

Fig. 29. *Lycoperdon*, this occurs with the early rain fall of rain. It can be eaten when it is young.

Fig. 30. Stages in the breaking of outer peridium of *Astraeus*

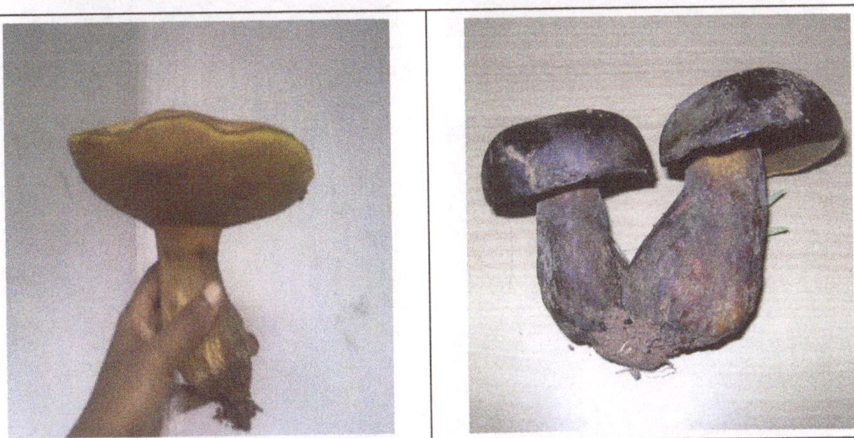

Fig. 31. *Boletus edulis* Bull. Always found associated with *Syzygium cuminii* Linn. This is the biggest mushroom found in Jharkhand.

Fig. 32. *Russula sanguinea* (Bull.) Fr. with millipede.

Fig. 33. Amanita *sp*.

Fig. 34. *Amanita* sp.	Fig. 35. *Amanita*, Collected from Ormanjhi

Fig. 36. *Clitocybe* collected from Manresa House Campus	Fig. 37. *Termitomyces microcarpous* (Var). this is small sized mushrooms, grows in the sandy soil therefore it is called *balu khukhri* (sand mushroom).

Fig. 38. *Armillaria mellea* **(Vahl) P. Kummgrowing under the** *quiscalis indica* **plant. It grows in group.**

Fig. 39. *Termitomyces fuliginosus*

Fig. 40. Name not known

Fig. 41. *Dictyophora*

Fig. 42. Name not known

Fig. 43. Mushroom collected from Mango tree

Some Unidentified Mushrooms

Literature Refered

Abe, T., Matsumoto, T. 1979. Studies on the distribution and ecological role of termites in a lowland rain forest of west Malaysia. Distribution and abundance of termites in Pasoh Forest Reserve. *Japanese Journal of Ecology,* 29:337–351.

Adhikari, M. K., Devkota, S. and Tiwari, R.D. 2005. Ethnomycological knowledge and uses of wild mushrooms in western and central Nepal. *Our Nature,* 3:13-19.

Adhikari, M. K.; Adhikari, K.S.; Joshi, A.R.; Bhandary, R. C., Gyawali, M. B. and Pradhan, B. 1996. *Studies on food value and mycotoxic wild mushrooms of Nepal.* Submitted to APINMAP/SCAPMAP/UNESCO committee of Nepal, Ministry of education, Nepal. p. 92.

Adhikari, M.K. 1988. Some higher fungi from Lang tang and its adjoining area. *In: Proceedings of National Conference on Science and Technology* organized by Royal Nepal Academy of Science and Technology (RONAST), Kathmandu: 237-241.

Ainsworth, G.C. 1973. *Introduction and Keys to Higher Taxa. In: The Fungi: An advance Treatise.*Vol. IV. Academic Press, New York, 1-7.

Alasoadura, S.O. 1966. Studies in the higher fungi of Nigeria, Macro fungi associated with termites. *Nova Hedwigia* 11: 387-383.

Apetorgbor, M M., Apetorgbor, A.K. and Nutakor, E. 2005. Utilization and cultivation of edible Mushrooms for rural livelihood in Southern Ghana. *17th Commonwealth Forestry Conference,* Colombo, Sri Lanka.

Bahl, Nita. 2005. *Handbook on Mushrooms.* New Delhi: Oxford & IBH Publishing CO.PVT.LTD, 28-29.

Batra, L.R., Batra, S.W.T. 1979. *Termite fungus mutualism. In Insect Fungus Symbiosis* (ed.L.R. Batra), New York, U.S.A.: John Willey and Sons, pp. 16-163.

Beausejour, Terri Marie. 1999. Getting started with mushroom cultivation. *Mushroom: the Journal of Wild Mushrooming*

Bhandary, H.R. 1985. Mushrooms. *In: Nepal Nature's Paradise* (Ed.Majpuria, T. C) Craftman press, Bangkok, Thailand.

Botha, W.J. and Eicker, A. 1968. Nutritional value of *Termitomyces* mycelial protein and growth of mycelium on natural substrates,

Buyck, B. 1994. Ubwoba: Les champignons comestibles de I'Ouest du Burundi. *Administ. generale de la cooperation au development Rue du Trone,* 4-1050 Bruxelles. Publication Agricole No 34.

Chan, H. K. M. 1981. Consumption of edible mushrooms in Hong Kong. *MushroomNewsletter for the Tropics* 1(4): 5-10.

Chang, S.T., Quimio, T.H. 1982. *Tropical Mushrooms Biological Nature and Cultivation Methods.* The Chinese University Press of Hong Kong. pp. 445-460.

Collins, N.M. 1983. The utilization of nitrogen resources by termites (Isoptera), In. Lee JA, McNeill S, Rorison IH (ed.), Nitrogen as an ecological factor, *Blackwell Scientific,* Oxford, United Kingdom. p. 381-412.

Cooke, R.C. 1977. *Fungi, man and his environment.* Longman group Ltd., London.

Cotter, H.V.T. a nd Miller, O.K 1987. Ectomycorrhizal association of the *Bolete* genus *Suillius* in Nepal. *In: Proceeding of Seventh North American conference in mycorrhizae,* Ganeaville, Fl.U.SA: 5.Department of Botany, University of Pretoria, Pretoria 0001, South Africa, Available online 11 February 2009.

Deacon, J.W. 1984. *Introduction to Modern Mycology.* Blackwell Scientific Publications, Oxford, 2nd edition, UK, 199.

Devkota, S. and Yarsagumba. 2006. Traditional utilization in Dolpa district, Western Nepal. *Our Nature.* (An International Biological Journal). 4:48-52.

Donatha Damian Tibuhwa, 2012. Termitomyces Species from Tanzania, Their Cultural Properties and Unequalled Basidiospores, *Journal of Biology and Life Science,* ISSN 2157-6076, Vol. 3, No. 1

Dutta, Ram. 2007. *Advances in Mushrooms Science.* Delhi: Satish Serial Publishing House, 24.

Gbolagade Jonathan.2006. Nutritive Value of Common Wild Edible Mushrooms from Southern Nigeria, *Global Journal of Biotechnology & Biochemistry,* 1 (1) 16-21.

Giri, A. and Rana, P. 2008. Ethonomycological Knowledge and Nutritional Analysis of Some Wild Edible Mushrooms of Sagarmatha National Park (SNP), Nepal, *J. Nat. Hist. Mus.* Vol. 23, 2008, 65-67.

Gogoi Robin, Rathaiah, Yella and Borah, Tasvina Rahman. 2006 *Mushroom Cultivation Technology.* 9.

Gogoi, R. R.Y. and B.T.R, *Mushroom Cultivation Technology,* (Scientific Publishers, Jodhpur), 2006, 111-112.

Gogoi, Robin. Rathaiah, Yella and Borah, Tasvina Rahman. 2006. *Mushroom Cultivation Technology*. Scientific Publishers India, Jodhpur, 4.

Gómes, L.D. 1995. Una nueva specie Neotropical de *Termitomyces (Agricales:Termitomycetaceae)*. *Revista de biologia Tropical* 42: 439-451.

Gour Pathak Yadav. 2010. *Mushroom Production and Processing Technology*. Jodhpur: Agrobios, 60.

Grassé, P.P., Noirot, C. 1958. Le meule des termites champignonnistes et sa signification symbiotique. *Ann. Sci. Nat. Zool. Biol. Anim.* 20:113-128.

Härkönen, M.T., Niemelä, T. and Mwasumbi, L. 2003. Tanzanian Mushrooms Edible, harmful and other fungi. *Botanical Museum Finnish Museum of Natural History,* 200.

Härkönen, M.T., Saarimaki and Mwasumbi, L. 1995. *Edible mushroom of Tanzania*. KARSTENIA, 35: Suppl., 92.

Heim, R. 1942. Nouvelles e'tudes descriptives sur les Agarics termitophiles d'Afrique tropicale. *Archives du Muse'um National d'Histoire Naturelle se'r.* 6, 18, 107-166.

Heim, R. 1977. *Termites et Champignons.Les termitophiles d'Afrique Noire at d'Asie Meridionale*. Paris, France: Societe Novelle des Edition. 205.

Henrik, H., De Fine Licht, Anders Andersen and Duur K. Aanen. 2005. Termitomyces sp. associated with the termite Macrotermes natalensis has a heterothallic mating system and multinucleate cells, *Mycol.* Res. 109 (3): 314–318 (March 2005). The British Mycological Society doi: 10.1017/S0953756204001844 Printed in the United Kingdom.

Herald Express (Torquay UK) March 6, 2008

Hobbs, C. 1987. *Medical mushrooms*. Botanica press, Box 742 Capitola, Ca 95010.

Johnsy, G. S., Davidson Sargunam, Dinesh, M.G. and Kaviyarasan,V. 2011. Nutritive Value of Edible Wild Mushrooms Collected from the Western Ghats of Kanyakumari District, *Botany Research International,* 4 (4) 69-74.

Kaul, T.N. 1999. *Introduction to mushroom Science*. Oxford & IBH Publishing Co.PVT.LTD, New Delhi, 30-32.

Kaul, T.N. 2002. *Biology and Conservation of Mushrooms*. Oxford & IBH Publishing CO.PVT.LTD, New Delhi, 3.

Kharel, S., and Rajbhandary, S. 2005. Ethno mycological knowledge of some wild edible mushrooms in Bhardeo, Lalitpur, Nepal. *Nepal Journal of Plant Sciences,* 1:45-49.

Li, G.S.F. and Chang, S.T. 1982. Nutritive value of Volvariella volvacea. *In: Tropical Mushroom. Biological nature and cultivation methods*. Ed. Chang.

Louis C.C. Krieger, 2007. *A Guide to the Mushrooms, their Botanical Position, Mode of growth, Structure, Physiology, Habitat, Ecology, and Economic Importance*. Asiatic Publishing House, Delhi.203-222.

Martin, M.M., Martin, J.S. 1978. Cellulose digestion in the midgut of the fungus-growing termites *Macrotermes natalensis*: the role of acquired digestive enzymes. *Science* 199:1453-1455. http://dx.doi.org/10.1126/science.199.4336.1453.

Michael Allaby. "Geastrum." A Dictionary of Plant Sciences. 1998. *Encyclopedia.com*. 16 Apr. 2010

Morris, B. 1986. Notes on the genus *Termitomyces* Heim in Malawi. *Society of Malawi Journal* 39, 40-49.

Nandkumar M Kamat. 2011. The Neglected Mushrooms of Goa Published on: *Panorama* July 16, 2011 - 23:22.

Natarajan, N.K. 1979. South Indian Agaricales v: *Termitomyces heimii. Mycologia*, 71, 853-855.http://dx.doi.org/10.2307/3759201.

Ogundana, S.K. and Fagade, O.E. 1982. Nutritional Value of Some Nigerian Edible Mushrooms. *Food Chem*. 8:263-268.

Otieno, N.C. 1964. Contributions to knowledge of Termite fungi in East Africa. *Proceedings of the East African Academy* 11, 108-120.

Pandey, N. 2006. *Preliminary study on the mushrooms diversity of Lang tang National Park*. A report submitted to the Department of National park and Wildlife Conservation, Babarmahal, Kathmandu. 19.

Pandey, N. and Budhathoki, U. 2007. Protein determination through Bradford's method of Nepalese mushroom. *Scientific world*, 5(5):85-88.

Pandey, R.K. and. Ghosh, S.K. 1999. *A Hand Book on Mushrooms Cultivation*. Emkay Publications, New Delhi, 9.

Pegler, D.N. 1977. A preliminary Agaric flora of East Africa. *Kew Bull. Add. Ser* 6: 1-615. London. HMSO.

Petch, T. 1913. Termite fungi: a résumé. *Ann Roy Bot Gard* (Peradeniya) 5:303–341.

Rana, P. and Giri, A. 2006. Diversity of mushroom in Sagarmatha

National Park, Nepal. *Banko Janakari (A Nepalese Journal of forestry)*, 16(2):17-24.

Rohrmann, G.F. 1978. The origin, structure and nutritional importance of the comb in two species of Macrotermitinae. *Pedobiologia* 18:89-98.

Rouland-Lefèvre, C. 2000. Symbiosis with fungi, In Abe T, Bignell DE, Higashi M (ed.), Termites: evolution, society, symbioses, ecology, p. 289-306. Kluwer Academic Publishers, Dordrecht, *The Netherlands*.

S4/USAID/102. *The International Centre of Insect Physiology and Ecology* P.O. Box 30772, Nairobi, Kenya Your Ref: Grant No. 936-5543-G-00-4023-00

Sacherer, J. 1979. The high altitude ethno botany of the Rowaling Sherpas: A contribution of Nepalese studies. *Journal of Research in Central Nepal Studies*, 6(2):46-64.

Singh, S. C. and Nisha. 1973. Experimental cultivation of *Pleurotus* in Nepal. *Jour. Trib. Univ,* 8:81-85.

Staben, C. 1994 .Ssexual reproduction in higher fungi. In Gow, N.A.R. and Gadd, G.M. (eds*) The growing fungus*. Chapman and Hall, London, 383-402.

Stamets, P. and Chilton, J. 1983. *The Mushroom Cultivator: A Practical Guide to Growing Mushrooms at Home*. Agarikon Press, Washington.

Suman, B.C. and Sharma, V.P. 2007. *Mushroom Cultivation in India*. Delhi: Daya Publishing House, 25.

Tanti Bhaben, Gurung Lisha & Sarma Gajen Chandra. 2011. Wild edible fungal resources used by ethnic tribes of Nagaland, India, *Indian Journal of Traditional Knowledge*, Vol. 10 (3), July 2011, pp. 512-515.

Toru Johjima, Yaovapa Taprab, Napavarn Noparatnaraporn, Toshiaki Kudo and Moriya Ohkuma. 2006.Large-scale identification of transcripts expressed in a symbiotic fungus (Termitomyces) during plant biomass degradation, *Appl Microbiol Biotechnol* (2006) 73:195–203 DOI 10.1007/s00253-006-0570-8

Tripathi, D.P. 2005. *Mushrooms Cultivation*. New Delhi: Oxford & IBH Publishing CO.PVT.LTD, 200.

Vander Westhuisen, G.C.A. and Eicker, and A. 1990. Species of *Termitomyces* in South Africa *Mycological Research,* 94: 923-937.

Wasson, R.G. 1968. *SOMA-Divine Mushrooms of Immortality*. Harcourt Javanovich Inc.

Wikipedia

William John S. ed. 1997. *Management of Natural Wealth*. Enviro Club of Loyola College, Chennai, 63.

Wong, W, C. and Cheung, P.C.K. 2001. *Food and Nutritional Science Programme*, The Chinese university of Hong Kong, Shatin, New territories, Hong Kong, China.

Wood, T.G., Sands, W.A. 1978. The role of termites in ecosystems. In: Brian MV (ed) *Production ecology of ants and termites.* Cambridge University Press, Cambridge, pp 245–292.

www.botany.hawaii.edu/faculty/wong/BOT135/Lect17a.html, *The Wild Edible Mushrooms.*

Yamada, A., Inoue, T., Wiwatwitaya, D., Ohkuma. M., Kudo, T., Abe, T.and Sugimoto, A. 2005. Carbon mineralization by termites in tropical forests, with emphasis on fungus-combs. *Ecol Res* 20:453–460.

Yang, Z.L. 2002. On wild mushroom resources and their utilization in Yunnan province, south-western China. *Journal of Natural Resources,* 17:464-469.

Zhang, Z.F., Ruan, X.Y. 1986. A new species *Termitomyces macrocarpus* Zhang and Ruan sp.nov. *Edible Fungi,* 5, 10-13.

Links of the Creative Comons Images Used

Name of the Image	Link of the Image	Website, Image Credit
Gill Fungi	https://www.rawpixel.com/image/4025179/photo-image-plant-nature	rawpixel.com
Pore Fngi	https://commons.wikimedia.org/wiki/File:Orange_pore_fungus_-_Favolaschia_calocera.jpg	commons.wikimedia.org
Teeth Fungi	https://commons.wikimedia.org/wiki/File:Hydnellum_aurantiacum,_known_as_the_Orange_Tooth_Fungus_%289618136186%29.jpg	commons.wikimedia.org
Club Or Coral Fungi	https://picryl.com/media/coral-fungi-clavaniaceae-feafd6	picryl.com
Jelly Fungi	https://commons.wikimedia.org/wiki/File:Goldgelber_Zitterling-Tremella_mesenterica_stack32_-20191227-RM-152807.jpg	commons.wikimedia.org
Dictyophora	https://pixabay.com/photos/dictyophora-indusiata-2731742/	pixabay.com
Ithyphallus	https://garystockbridge617.getarchive.net/amp/media/synopsis-of-the-british-basidiomycetes-a-descriptive-catalogue-of-the-drawings-40a2e3	garystockbridge617.getarchive.net
Geaster	https://www.flickr.com/photos/volvob12b/41915597414	www.flickr.com
Lycoperdon	https://commons.wikimedia.org/wiki/File:Lycoperdon_pyriforme_%28Pear_shaped_puffball%29.jpg	commons.wikimedia.org
Bovista	https://commons.wikimedia.org/wiki/File:Bovista_aestivalis.jpg	commons.wikimedia.org
Calvatia	https://commons.wikimedia.org/wiki/File:Calvatia_craniiformis.JPG	commons.wikimedia.org
Cyathus	https://commons.wikimedia.org/wiki/File:Bird%27s_Nest_Fungus_%28Cyathus_olla%29_-_Flickr_-_Andrea_Westmoreland.jpg	commons.wikimedia.org
Morchella	https://commons.wikimedia.org/wiki/File:Morchella_esculenta_84915.jpg	commons.wikimedia.org
Gyromitra	https://commons.wikimedia.org/wiki/File:Gyromitra_esculenta_%2847348478352%29.jpg	commons.wikimedia.org
Helvella	https://commons.wikimedia.org/wiki/File:Helvella_dryophila_597531.jpg	commons.wikimedia.org
Plicaria	https://commons.wikimedia.org/wiki/File:Plicaria_endocarpoides_13215.jpg	commons.wikimedia.org

Name of the Image	Link of the Image	Website, Image Credit
Sarcoscypha	https://commons.wikimedia.org/wiki/File:Sarcoscypha_coccinea_%28Scop.%29_%282%29.jpg	commons.wikimedia.org
Geopyxis	https://commons.wikimedia.org/wiki/File:Geopyxis_carbonaria_%288755670533%29.jpg	commons.wikimedia.org
Cup Fungi, Pizza Sp	https://pxhere.com/en/photo/1605375	pxhere.com
Morels (Morchella)	https://commons.wikimedia.org/wiki/File:Morchella_esculenta_84915.jpg	commons.wikimedia.org
Lepista nuda	https://commons.wikimedia.org/wiki/File:Lepista_nuda_%2827105376950%29.jpg	commons.wikimedia.org
Volvariella	https://commons.wikimedia.org/wiki/File:Volvariella_bombycina_427539.jpg	commons.wikimedia.org
Marasmium	https://commons.wikimedia.org/wiki/File:Marasmius_oreades_garden_050829B.JPG	commons.wikimedia.org
Polyporos tuberaster	https://commons.wikimedia.org/wiki/File:Polyporus_tuberaster_%2836084108394%29.jpg	commons.wikimedia.org
Boletus	https://www.rawpixel.com/image/4023366/flammulina-velutipes	www.rawpixel.com
Lacterius	https://commons.wikimedia.org/wiki/File:Lactarius_rubrilacteus_126762.jpg	commons.wikimedia.org
Tricholoma	https://commons.wikimedia.org/wiki/File:Tricholoma_pardinum_gills.jpg	commons.wikimedia.org
Termitomyces	https://commons.wikimedia.org/wiki/File:Termitomyces_microcarpus_%28Berk._%26_Broome%29_R._Heim_564259.jpg	commons.wikimedia.org
Agrocybe	https://commons.wikimedia.org/wiki/File:Agrocybe_praecox_20120415wa.JPG	commons.wikimedia.org
Pleurotus	https://commons.m.wikimedia.org/wiki/File:20210719_Pleurotus_ostreatus_001-016.jpg	commons.wikimedia.org
Auricularia	https://commons.wikimedia.org/wiki/File:Auricularia_fuscosuccinea_419264.jpg	commons.wikimedia.org
Lentinus	https://commons.wikimedia.org/wiki/File:Lentinus_sp_7871.jpg	commons.wikimedia.org
Armellariella	https://commons.wikimedia.org/wiki/File:Honey_Fungus_%28Armillaria_mellea%29.jpg	commons.wikimedia.org
Cyttaria	https://commons.wikimedia.org/wiki/File:Cyttaria_espinosae.jpg	commons.wikimedia.org
Agricus	https://www.publicdomainpictures.net/en/view-image.php?image=178430&picture=mushrooms	www.publicdomainpictures.net

Name of the Image	Link of the Image	Website, Image Credit
Caprinus	https://commons.wikimedia.org/wiki/File:Coprinus_comatus_%28GB%3D_Shaggy_Inkcap_Syn._Lawyer%27s_Wig,_D%3D_Schopftintling,_Spargelpilz,_F%3D_Coprin_chevelu,_NL%3D_Geschubde_inktzwam%29_in_3_growing_stadia_at_Bennekom_-_panoramio.jpg	commons.wikimedia.org
Amanita	https://commons.wikimedia.org/wiki/File:2006-10-25_Amanita_muscaria_crop.jpg	commons.wikimedia.org
Cartharellus	https://commons.wikimedia.org/wiki/File:2007-07-14_Cantharellus_cibarius2.jpg	commons.wikimedia.org
Clitocybe	https://commons.wikimedia.org/wiki/File:Clitocybe_nuda_%28Fr.%29_H.E._Bigelow_%26_A.H._Sm_267650.jpg	commons.wikimedia.org
Lacterius	https://commons.wikimedia.org/wiki/File:Bruderwald_Reitzker_%28Lactarius_deliciosus%29_-RM-20191129-01.jpg	commons.wikimedia.org
Lentinus	https://commons.wikimedia.org/wiki/File:2012-12-31_Lentinus_strigosus_%28Schwein.%29_Fr_299086.jpg	commons.wikimedia.org
Lepiota	https://commons.wikimedia.org/wiki/File:2011-12-01_Lepiota_cristata_%28Bolton%29_P._Kumm_186612.jpg	commons.wikimedia.org
Pleurotus	https://commons.wikimedia.org/wiki/File:Oyster_Mushroom_%28Pleurotus_ostreatus%29_%2816268204621%29.jpg	commons.wikimedia.org
Russulla	https://commons.wikimedia.org/wiki/File:Russula_silvicola_299912.jpg	commons.wikimedia.org
Tricholoma	https://commons.wikimedia.org/wiki/File:Tricholoma_griseoviolaceum_34304.jpg	commons.wikimedia.org
Flammulina	https://commons.wikimedia.org/wiki/File:Flammulina_velutipes_IMG_1038.jpg	commons.wikimedia.org
Inocybe	https://commons.wikimedia.org/wiki/File:Inocybe_fuscidula_616149.jpg	commons.wikimedia.org
Paxillus	https://commons.wikimedia.org/wiki/File:Paxillus_involutus_EtgHollande_041031_063.jpg	commons.wikimedia.org
Entoloma	https://commons.wikimedia.org/wiki/File:2009-06-27_Entoloma_griseocyaneum.jpg	commons.wikimedia.org
Volvariella	https://commons.wikimedia.org/wiki/File:Volvariella_pusilla_130922wb.JPG	commons.wikimedia.org
Agaricus	https://commons.wikimedia.org/wiki/File:Agaricus-campestris-michoacan.jpg	commons.wikimedia.org
Hypholoma	https://commons.wikimedia.org/wiki/File:Hypholoma_fasciculare_LC0091.jpg	commons.wikimedia.org

Name of the Image	Link of the Image	Website, Image Credit
Stopharia	https://commons.wikimedia.org/wiki/File:2011-05-19_Stropharia_rugosoannulata_Farl._ex_Murrill_183478.jpg	commons.wikimedia.org
Caprinus	https://commons.wikimedia.org/wiki/File:Coprinus_Comatus_02.jpg	commons.wikimedia.org
panaelus	https://commons.wikimedia.org/wiki/File:Panaeolus_papilionaceus_California.jpg	commons.wikimedia.org
Psathyrella	https://commons.wikimedia.org/wiki/File:Psathyrella_longipes.jpg	commons.wikimedia.org
Amanita	https://commons.wikimedia.org/wiki/File:2008-10_Amanita_pantherina_cropped.jpg	commons.wikimedia.org
Favolus	https://commons.wikimedia.org/wiki/File:Favolus_tenuiculus_-_Flickr_-_Dick_Culbert.jpg	commons.wikimedia.org
Hydnum	https://commons.wikimedia.org/wiki/File:Hydnum_umbilicatum_Peck_769754.jpg	commons.wikimedia.org
Boletinus	https://commons.wikimedia.org/wiki/File:Boletus_cavipes.jpg	commons.wikimedia.org
Clavaria	https://commons.wikimedia.org/wiki/File:Clavariadelphus-pistillaris.jpg	commons.wikimedia.org
Exedia	https://commons.wikimedia.org/wiki/File:2011-11-12_Exidia_thuretiana_181923.jpg	commons.wikimedia.org
Tremella	https://commons.wikimedia.org/wiki/File:Goldgelber_Zitterling-Tremella_mesenterica_stack32_-20191227-RM-152807.jpg	commons.wikimedia.org
Xylaria	https://commons.wikimedia.org/wiki/File:Xylaria_hypoxylon_%28GB%3D_Candlestickfungus_Syn._Carbon_antlers,_D%3D_Geweihf%C3%B6rmige_Holzkeule,_F%3D_Xylaire_du_bois,_NL%3D_Geweizwam%29_white_conidium_asexual_spores,_later_black_spores_and_causes_white_root_rot,_at_Arnhem_Water_-_panoramio.jpg	commons.wikimedia.org
Calvatia	https://commons.wikimedia.org/wiki/File:Calvatia_Fr_620036.jpg	commons.wikimedia.org

Index

J

K

L

M

N

O

Colour Plates

Gill fungi (P. 7)

Pore fungi (P. 7)

Teeth fungi (P. 7)

Club or Coral (P. 8)

Jelly fungi (P. 8)

Dictyophora (P. 8)

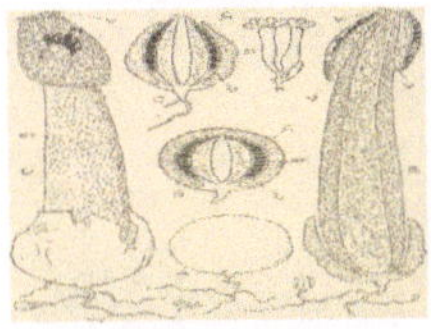

Ithyphallus garystock-bridge 617 (P. 8)

Geaster (P. 8)

Lycoperdon (P. 8)

Bovista (P. 8)

Calvatia (P. 8)

Cyathus (P. 8)

Morchella (P. 8)

Gyromitra (P. 8)

Helvella (P. 8)

Plicaria (P. 8)

Sarcoscypha (P. 8)

Geopyxis (P. 8)

Cup Fungi, *Pizza* sp. (P. 9)

Morel, *Morchella* sp. (P. 9)

Lepista nuda (P. 9)

Volvariella (P. 9)

Marasmium (P. 10)

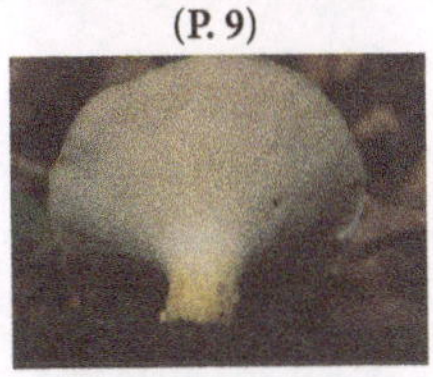

Polyporus tuberaster (P. 10)

Boletus (P. 10)

Lacterius (P. 10)

Tricholoma (P. 10)

Termitomycetes (P. 10)

Agrocybe (P. 10)

Pleurotus (P. 10)

Auricularia (P. 11)

Lentinus (P. 11)

Armellariella mellea (P. 11)

Cyttaria (P. 11)

Agaricus (P. 11)

Caprinus (P. 11)

Amanita (P. 13)

Cantharellus (P. 13)

Clitocybe (P. 13)

Lacterius (P. 13)

Lentinus (P. 13)

Lepiota (P. 13)

Pleurotus (P. 14)

Russula (P. 14)

Tricholoma (P. 14)

Flammulina (P. 14)

Inocybe (P. 14)

Paxillus (P. 14)

Entoloma (P. 14)

Volvariella (P. 14)

Agaricus (P. 14)

Hypholoma (P. 14)

Stropharia (P. 14)

Caprinus (P. 14)

Panaeolus (P. 14)

Psathyrella (P. 14)

Amanita pantherina (P. 15)

Favolus tenuiculus (P. 15)

Hydnum umblicatum (P. 15)

Boletinus cavipes (P. 16)

Clavariadelphus pistillaris (P. 16)

Exidia thuretiana (P. 16)

Tremella (P. 16)

Xylaria hypoxylon (P. 16)

Calvatia (P. 16)

Fully grown Oyster mushroom (P. 23)

Volvariella (P. 25)

Polythene bags (P. 29)

Agaricus bisporus (P. 30)

Fig. 18 (P. 48)

Fig.19 (P. 49)

Fig. 20 (P. 49)

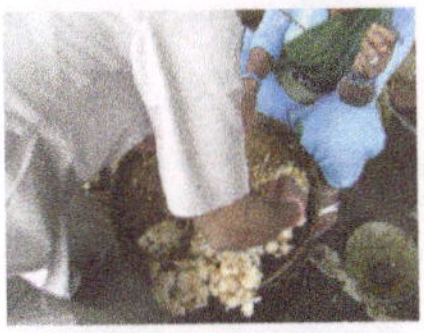
Fig. 21 (P. 49)

Fig. 22 (P. 49)

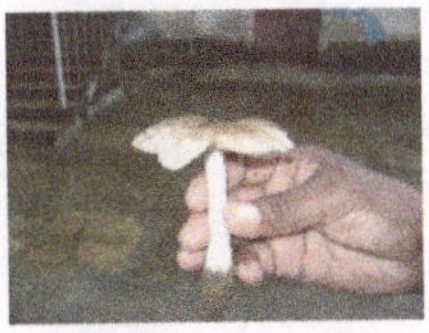
Fig. 23 (P. 49)

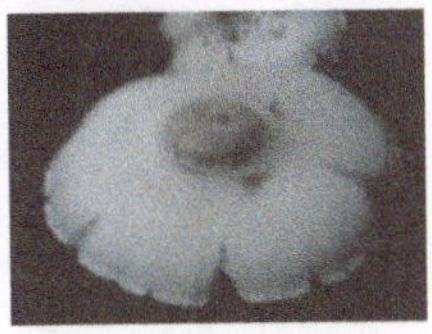
Fig. 24 (P. 49)

Fig. 25 Stages in the growth of *Bada Khukhri* (P. 50)

Fig. 26 (P. 50)

Fig. 27 (P. 50)

Fig.28 (P. 50)

Fig. 29 (P. 50)

IV

Fig. 30 (P. 51)

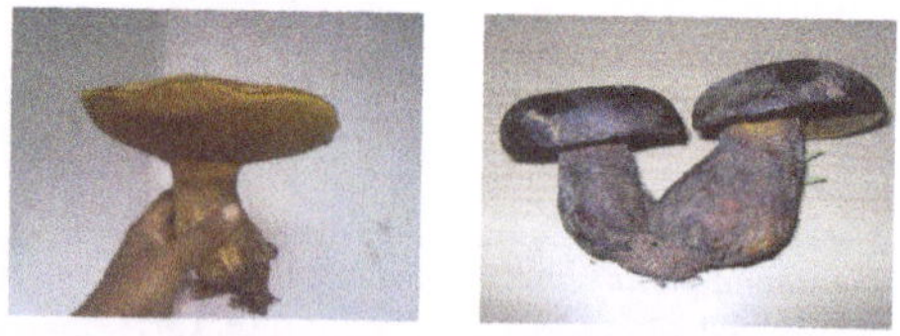

Fig. 31 (P. 51)

Fig. 32 (P. 51)

Fig. 33 (P. 51)

Fig. 34 (P. 52)

Fig. 35 (P. 52)

Fig. 36 (P. 52)

Fig. 37 (P. 52)

Fig. 38 (P. 52)

Fig. 39 (P. 53)

Fig. 40 (P. 53)

Fig. 41 (P. 53)

Fig. 42 (P. 53)

Fig. 43 (P. 53)

Fig. 43 (P. 53)

Some Unidentified Mushrooms (P. 54)